Geometric Modeling of Fractal Forms for CAD

Geometric Modeling and Applications Set

coordinated by
Marc Daniel

Volume 5

Geometric Modeling of Fractal Forms for CAD

Christian Gentil
Gilles Gouaty
Dmitry Sokolov

WILEY

First published 2021 in Great Britain and the United States by ISTE Ltd and John Wiley & Sons, Inc.

ISTE Ltd
27-37 St George's Road
London SW19 4EU
UK

www.iste.co.uk

John Wiley & Sons, Inc.
111 River Street
Hoboken, NJ 07030
USA

www.wiley.com

Library of Congress Control Number: 2021932086

British Library Cataloguing-in-Publication Data
A CIP record for this book is available from the British Library
ISBN 978-1-78630-040-9

Contents

Preface

This work introduces a model of geometric representation for describing and manipulating complex non-standard shapes such as rough surfaces or porous volumes. It is aimed at students in scientific education (mathematicians, computer scientists, physicists, etc.), engineers, researchers or anyone familiar with the mathematical concepts addressed at early stages of the graduate level. However, many parts are accessible to all, in particular, all introductory sections that present ideas with examples. People with no prior background, whether they are artists, designers or simply curious, will be able to understand the philosophy of our approach, and discover a new universe of unsuspected and exciting forms.

Geometric representation models are mathematical tools integrated into computer-aided geometric design (CAGD) software. They make the production of numerical representations of forms possible. By means of graphical interfaces or programming tools, users can draw and/or manipulate these shapes. They can also test or evaluate their physical properties (mechanical, electro-magnetic, acoustic, etc.) by communicating geometric descriptions to further specific numerical simulation software.

The geometric representation model we present here is based on the fractal geometry paradigm. The principle behind this consists of studying the properties (signal, geometry, phenomena, etc.) at different scales and identifying the invariants from there. The objects are described as self-referential between two scales: each of the object features (namely, the lower scale level) is described as a reference to the object itself (namely, the higher scale). This approach is not conventional and often confusing at first. We come to perceive its richness and power very quickly, however. The universe of forms that can possibly be created is infinite and has still only partially been explored.

In this book, we present the mathematical foundations, so that the reader can access all the information to understand, test and make use of this model. Properties, theorems and construction methods are supplemented with algorithms and numerous examples and illustrations. Concerning the formalization, we have chosen to use precise and rigorous mathematical notations to remove any ambiguity and make understanding easier.

Readers unwilling to be concerned with mathematical formalisms can get to grips with the philosophy of our approach by focusing on the sections found at the beginnings of the chapters, in which ideas and principles are intuitively presented, based on examples.

This book is the result of 25 years of research carried out mainly in the LIRIS laboratories of the University of Lyon I and LIB of the University of Burgundy Franche-Comté. This research was initiated by Eric Tosan, who was instrumental at the origin of this formalism and to whom we dedicate this work.

Christian GENTIL
Gilles GOUATY
Dmitry SOKOLOV

February 2021

Introduction

I.1. Fractals for industry: what for?

This book shows our first steps toward the fundamental and applied aspects of geometric modeling. This area of research addresses the acquisition, analysis and optimization of the numerical representation of 3D objects.

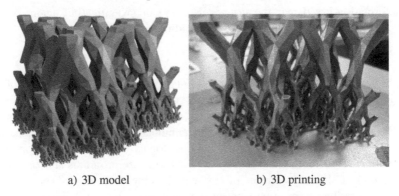

a) 3D model   b) 3D printing

Figure I.1. *3D tree built by iterative modeling (source: project MODITERE no. ANR-09-COSI-014)*

Figure I.1 shows an example of a structure that admits high vertical loads, while minimizing the transfer of heat between the top and bottom of the part. Additive manufacturing (3D printing) allows, for the first time, the creation of such complex objects, even in metal (here with a high-end laser printer EOS M270). This type of technology will have a high societal and economic impact, enabling better systems to be created (engines, cars, airplanes, etc.), designed and adapted numerically for optimal functionality, thus consuming less raw material, for their manufacturing, and energy, when used.

Current computer-aided design is, however, not well suited to the generation of such types of objects. For centuries, for millennia, humanity has produced goods with axes, files (or other sharp or planing tools), by removing bits from a piece of wood or plastic. Tools subsequently evolved into complex numerical milling machines. However, at no point during these manufacturing processes did we need sudden stops or permanent changes in the direction of the cutting tool. The patterns were always "regular", hence the development of mathematics specific to these problems and our excellent knowledge of the modeling of smooth objects. This is why it was necessary to wait until the 20th century to have the mathematical knowledge needed to model rough surfaces or porous structures: we were just not able to produce them earlier.

Thus, since the development of computers in the 1950s, computer-aided geometric design (CAGD or CAD) software has been developed to represent geometric shapes intended to be manufactured by standard manufacturing processes. These processes are as follows:

– subtractive manufacturing, using machine tools such as lathes or milling machines;

– molding, where molds themselves are made using machine tools;

– deformation-based manufacturing: stamping or swaging (but again, dies are usually manufactured using machine tools), folding, etc.;

– cutting, etc.

Each of these processes imposes constraints, for example, concerning collision issues in milling machines (even a five-axis mill cannot produce any geometry). These manufacturing processes inevitably influenced the way we design the geometries of objects, in order to actually manufacture them. For example, CAD software has integrated these design methodologies by developing appropriate numerical models or tools. Currently, most CAD software programs are based on the representation of shapes by means of surfaces defining their edges. These surfaces are usually described using a parametric representation called non-uniform rational B-spline (NURBS). These surface models are very powerful and very practical. It is possible to represent any volume enclosed by a quadric (cylinders, cones, spheres, etc.) and complex shapes, such as car bodies or airplane wings. They were originally designed for this.

However, the emergence of additive manufacturing techniques has caused an upheaval in this area, opening up the possibility of potentially "manufacturable" forms. By removing the footprint constraint of the tool, it then becomes possible to produce very complex shapes with gaps or porosity. These new techniques have called into question the way objects are designed. New types of objects, such as porous objects or rough surfaces, can have many advantages, due to their specific physical properties. Fractal structures are used in numerous fields such as architecture (Rian and Sassone 2014), jewelry (Soo *et al.* 2006), heat and mass transport (Pence 2010), antennas (Puente *et al.* 1996; Cohen 1997) and acoustic absorption (Sapoval *et al.* 1997).

I.2. Fractals for industry: how?

The emergence of techniques such as 3D printers allows for new possibilities that are not yet used or are even unexplored. Different mathematical models and algorithms have been developed to generate fractals. We can categorize them into three families, as follows:

– the first groups algorithms for calculating the attraction basins of a given function. Julia and Mandelbrot (Peitgen and Richter 1986) or the Mandelbulb (Aron 2009) sets are just a few examples;

– the second is based on the simulation of phenomena such as percolation or diffusion (Falconer 1990);

– the last corresponds to deterministic or probabilistic algorithms or models based on the self-similarity property associated with fractals such as the terrain generator (Zhou *et al.* 2007), the iterated function system (Barnsley *et al.* 2008) or the L-system (Prusinkiewicz and Lindenmayer 1990).

In the latter family of methods, shapes are generated from rewriting rules, making it possible to control the geometry. Nevertheless, most of these models have been developed for image synthesis, with no concerns for "manufacturability", or have been developed for very specific applications, such as wood modeling (Terraz *et al.* 2009). Some studies approach this aspect for applications specific to 3D printers (Soo *et al.* 2006). In (Barnsley and Vince 2013b), Barnsley defines fractal homeomorphisms of $[0, 1]^2$ onto the modeling space $[0, 1]^2$. The same approach is used in 3D to build 3D fractals. A standard 3D object is integrated into $[0, 1]^3$ and then transformed into a 3D fractal object. This approach preserves the topology of the original object, which is an important point for "manufacturability".

The main difficulty associated with traditional methods for generating fractals lies in controlling the forms. For example, it is difficult to obtain the desired shape using the fractal homeomorphism system proposed by Barnsley. Here, we develop a modeling system of a new type based on the principles of existing CAD software, while expanding their capabilities and areas of application. This new modeling system offers designers (engineers in industry) and creators (visual artists, designers, architects, etc.) new opportunities to quickly design and produce a model, prototype or unique object. Our approach consists of expanding the possibilities of a standard CAD system by including fractal shapes, while preserving ease of use for end users.

We propose a formalism based on standard iterated function systems (IFS) enhanced by the concept of boundary representation (B-rep). This makes it possible to separate the topology of the final forms from the geometric texture, which greatly simplifies the design process. This approach is powerful, and it generalizes subdivision curves and standard surfaces (linear, stationary), allowing for additional control. For example, we have been able to propose a method for connecting a primal subdivision scheme surface with a dual subdivision scheme surface (Podkorytov *et al.* 2014), which is a difficult subject for the standard subdivision approach.

The first chapter recalls the notion of self-similarity, intimately linked to that of fractality. We present the IFS, formalizing this property of self-similarity. We then introduce enhancements into this model: controlled iterated function systems (C-IFS) and boundary controlled iterated function systems (BC-IFS). The second chapter is devoted to examples. It provides an overview of the possibilities of description and modeling of BC-IFS, but also allows better understanding the principle of the model through examples. The third chapter presents the link between BC-IFS, the NURBS surface model and subdivision surfaces. The results presented in this chapter are important because they show that these surface models are specific cases of BC-IFS. This allows them to be manipulated with the same formalism and to make them interact by building, for example, junctions between two surfaces of any kind. In the fourth chapter, we outline design tools that facilitate the description process, as well as examples of the applications, of the design of porous volumes and rough surfaces.

1

The BC-IFS Model

In this chapter, in section 1.1, we begin by intuitively introducing the notion of self-similarity. Then, in section 1.1.2, we give its mathematical formulation as proposed by Hutchinson (Hutchinson 1981) using iterated function systems (IFS). Next, we present how this mathematical model can be implemented to calculate and visualize geometric shapes.

In section 1.2, we set out an extension of the IFS, allowing us to move away from strict self-similarity and generate a larger family of forms. This extension is called a *controlled iterated function system* (C-IFS).

Finally, the final step in formalization is to enhance the C-IFS model with the notion of boundary representation (B-rep). This step is fundamental because it will make it possible to describe and control the topology of fractal forms.

1.1. Self-similarity and IFS

In this book, when we talk about fractal shapes, fractal objects or simply fractals, it is in the sense of self-similar objects.

There are different definitions of self-similarity. For example, in the field of image processing, an image is considered self-similar when certain parts of that image are identical to (or "look alike") other parts. This property is exploited, in particular, by *inpainting*, a technique that consists of reconstructing part of a damaged or deliberately subtracted image (for example, to erase a character). The property of self-similarity is often verified with natural images, which is why *inpainting* algorithms are so successful. The principle is to look for an area similar to the area that needs to be filled in, and copy it over the missing area.

In fractal geometry, it has a different meaning.

DEFINITION.– *An object is called self-similar if it is composed of copies of itself.*

This definition is not rigorous and further clarification has to be provided. Nonetheless, it contains the main idea. Rather than describing the complex structure of an object, we describe its parts using a reference to the object itself. It is therefore a self-referential or recursive definition.

For example, we can describe the structure of a tree as being composed of several parts: its main branches (see Figure 1.1). Each of its parts can be considered as a smaller tree.

Figure 1.1. *Schematic illustration of self-similarity. The black tree can be seen as a composition of two trees (in green and red). For a color version of this figure, see www.iste.co.uk/gentil/geometric.zip*

In this example, we understand how a complex shape, such as that of a tree, can be simply described. Self-similarity provides information about the structure of the object with a different approach from that of standard geometric representations. By referring to the parts of the object, we study the details of the shape, in other words we perform a change of scale. Each detail is then described according to the object itself, namely as if the detail "looked like" a reduced version of the object. The tree is made up of branches. Each branch is defined as a tree. We can then apply thereto the definition of the tree: it is a composition of branches. This reasoning can then be indefinitely iterated.

From this definition and this introductory example, a number of questions immediately arise:

– Is this description relevant?

– Which objects are self-similar?

– What does it mean when a detail "looks like" the object?

– Is the shape of an object completely determined from its self-similarity?

– Knowing the self-similarity property, is it possible to reproduce its shape?

– Can an object possess different self-similarities?

In order to better understand the self-similarity property and provide an early answer to some of these questions, we are going to consider a second example. However this time, we start from an object whose shape is not *a priori* known, but its self-similarity property is. Assume that this object consists of five main parts, each of which is an exact copy of the object on a smaller scale.

The left part of Figure 1.2 shows each part, symbolized by an arbitrarily chosen shape, a square in this case. In this illustration, each part is not identical to the overall structure. To meet the definition of self-similarity, we need to replace each part with the overall composite structure of the five forms. We thus obtain the image in the center of Figure 1.2. However, by adding details to the details, we have also added details to the overall structure that were not initially present. There is always a discrepancy between the details of the overall form and the details of each part. By iterating this construction to reduce this discrepancy, the same effect will then be observed. Consistency between the object and each of its parts can only be achieved if the process is applied an infinite number of times to obtain a result like the image on the right

Figure 1.2. *An example of a self-similar object composed of five copies of itself. These five main parts are symbolized on the left by the five squares. In the center, each square has been replaced by the overall form composed of the five parts. On the right-hand side, the same construction process has been applied seven times*

From the second example, the following observations can be derived:

– It shows how from the definition of self-similarity alone, a form can be built.

– It brings forward the recursive aspect of the definition.

– To obtain the final form, the construction process must be applied an infinite number of times. From a mathematical point of view, this is not a problem, but it is unfeasible from a practical point of view and in particular, in computer science.

– The shape originally chosen to represent each self-similar part (in this example, the square) is not relevant. Indeed, the details that we add are successively smaller. After a few iterations (seven or eight), they have little influence on the overall shape. When the details are smaller in size than the accuracy of the display medium or the accuracy of the manufacturing device, their initial shape has no influence on the result. Figure 1.3 shows the same self-similar structure, but is symbolized by the image of an elephant instead of a square. The replacement or substitution process remains the same. We can observe that we obtain an identical figure as soon as the details become very small. The final image is less dense than that obtained with the square, but the structure is the same.

– The definition of self-similarity that we propose is very inaccurate. What does "copy of the object" mean? For our example, the copies correspond to the object after applying geometric transformations: change of scale (<1), translation and rotation. These transformations are essential because they are the ones that tell us, with each iteration, how to replace each part with its details.

Figure 1.3. *The self-similarity property, as shown in Figure 1.2, is symbolized by means of a pink elephant instead of a square. For a color version of this figure, see www.iste.co.uk/gentil/geometric.zip*

COMMENTS ON FIGURE 1.3.– *The same construction process is used. In the right-hand side of the image, we observe that after seven iterations the resulting form is composed of very fine details: a set of very small elephants that we cannot discern. The overall shape, however, is identical to that shown in Figure 1.2.*

1.1.1. *Mathematical definitions and reminders*

The concept of self-similarity was formalized by Hutchinson, based on iterated function systems (IFS) (Hutchinson 1981). Subsequently, it was developed and popularized by Barnsley (1988). Because the construction of fractal objects is based on an iterative process, it is then appropriate to choose workspaces with "good properties" that ensure the convergence of these constructions. In this section, we give the definitions of the spaces from which a very general notion of objects is defined and which iterated systems of functions will be able to operate with.

We also recall the concept of contracting transformations and fixed-points. These two concepts justify the representation (or coding) of self-similar fractals by iterated function systems. They make it possible to show the existence and unicity of a fractal associated with an iterated system of functions.

This section is not essential in understanding the rest of this book. An intuitive understanding of the concept of contractive function and of the principle of the fixed-point theorem should be enough.

1.1.1.1. *Working space*

As a work environment, we choose a complete metric space, denoted by $(\mathbb{X}, d)$. The topology associated with the $\mathbb{X}$ space is simply that induced by the metric d. In practice, this space corresponds to the set in which fractal objects are designed. Generally, $\mathbb{X} = \mathbb{R}^2, \mathbb{R}^3$ and d correspond to the Euclidian distance, hereafter denoted by d_E. Subsequently, we shall see that $\mathbb{X}$ can designate a barycentric space. The latter are best suited to represent shapes according to control points.

A complete space is a space in which any Cauchy sequence converges. Intuitively, if the distance between the terms of a sequence can be as small as we wish, provided that the terms are chosen with sufficiently high indices, then we are guaranteed that this sequence possesses a limit in that same space.

In our space $(\mathbb{X}, d)$, an "object" is defined as a non-empty compact subset. The set of the objects is denoted by $\mathcal{H}(\mathbb{X})$. If space $\mathbb{X}$ is a finite-dimensional metric Space (which will be the case in our context), then a compact set is a closed and bounded set. A closed set is a set that contains its boundary. The fact that it is bounded means that it can be included inside a ball of radius $r < \infty$.

The definition that we give an object is very general and does not ensure that the object can be materialized or manufactured. It can have an empty interior and it can be composed of unconnected parts, or both. For example:

– if $(\mathbb{X}, d) = (\mathbb{R}^2, d_E)$, the set reduced to a point $\{(1, 2)\}$ is an object with an empty interior;

– if $(\mathbb{X}, d) = (\mathbb{R}, d_E)$, $\{x_n = \frac{1}{n}, n \in \mathbb{N}\} \cup \{0\}$ is an non-connected object with an empty interior.

It is helpful to associate a distance with the spaces that we work in. This allows us to make use of the topology induced by this distance and to simply manipulate the notion of neighborhood. We equip the set of objects with the Hausdorff distance, induced by that of the working space $(\mathbb{X}, d)$. This distance is defined as follows.

DEFINITION.– *Let* $(\mathbb{X}, d)$ *be a metric space, the Hausdorff distance between two elements* A *and* B *of* $\mathcal{H}(\mathbb{X})$ *is defined by:*

$$d_{\mathbb{X}}(A, B) = \max\{\sup_{a \in A} \inf_{b \in B} d(a, b), \sup_{b \in B} \inf_{a \in A} d(a, b)\}$$

The Hausdorff distance is a distance between sets. To intuitively understand this distance, consider two sets A and B. If the Hausdorff distance between these two sets is equal to ϵ, it means that for any point of A, a point B can be found at a distance less than or equal to ϵ. However, for any point of B, a point of A can also be found at a distance less than or equal to ϵ. In other words, if we take a ball of radius ϵ and we walk the center of that ball over all the points of A, the space covered by this ball will cover the set B. The same property must be verified when inverting the role of A and B (see Figure 1.4).

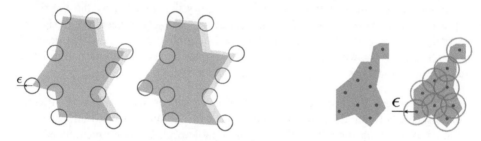

Figure 1.4. *Hausdorff distance. For a color version of this figure, see www.iste.co.uk/gentil/geometric.zip*

COMMENTS ON FIGURE 1.4.– *On the left, the Hausdorff distance between the two purple and green sets is less than or equal to* ϵ. *If the center of a disc of radius* ϵ *travels through all the points of the purple set, the set of discs will cover the green set. Conversely, if the center of a disc of radius* ϵ *travels through all the points of the green set, the set of discs will cover the purple set. On the right, the pink set is at a Hausdorff distance smaller than* ϵ *from the set consisting of blue dots. Indeed, each blue dot belongs to the pink set and the set of discs of radius* ϵ *centered at the blue dots covers the pink set. We observe that these last two sets do not share the same topological nature.*

We then define the set of objects in which we shall build our fractal objects using iterative processes.

DEFINITION.– *The space of the objects of* $(\mathbb{X}, d)$, *denoted* $\mathcal{H}(\mathbb{X})$, *is the set of the non-empty compact subsets of* $\mathbb{X}$ *endowed with the Hausdorff distance* $d_{\mathbb{X}}$, *associated with* d.

The following property is fundamental, it makes it possible to show that the mere knowledge of the self-similar property alone is sufficient to determine an object in a unique way.

PROPERTY.– If $(\mathbb{X}, d)$ is a complete metric space, then $(\mathcal{H}(\mathbb{X}), d_{\mathbb{X}})$ is a complete metric space.

This property also ensures the convergence of the iterative construction process of fractal objects (see section 1.1.4).

1.1.1.2. *Geometric transformations*

Geometric transformations play a major role in the formalization of the self-similarity property. As we have been able to understand through the introductory example (see Figure 1.2), they completely determine the geometry of the forms. Here, we recall some definitions and properties of transformations.

DEFINITION.– *A transformation is a function whose definition domain is equal to the starting space.*

DEFINITION.– *A transformation* $T : \mathbb{X} \to \mathbb{X}$, *with* $(\mathbb{X}, d)$ *as a metric space, is contractive if there is a real number* s, $0 \leqslant s < 1$, *such that* $d(T(x), T(y)) < s \cdot d(x, y) \; \forall x, y \in \mathbb{X}.$

The concept of contractive transformation appears implicitly in the self-similarity property when we wish to transform an object into one of its parts, which is necessarily smaller than the object itself. We shall see later that this condition is convenient but not necessary.

EXAMPLE.– Contractive transformations:

- $f : \mathbb{R} \to \mathbb{R}$ such that $f(x) = \frac{1}{2}x + \frac{1}{2}, (s = \frac{1}{2})$;
- $f : \mathbb{R}^2 \to \mathbb{R}^2$ such that $f(x, y) = (\frac{1}{3}x, \frac{1}{3}y)$, $(s = \frac{1}{3})$.

Here, we introduce a fundamental theorem on contractive transformations: the Banach fixed-point theorem. In our context, as we shall see hereafter, this theorem justifies the relevance of the definition of self-similarity by proving the existence and unicity of the object associated with a given self-similarity.

THEOREM.– Let $(\mathbb{X}, d)$ be a complete metric space and T a contractive transformation on $(\mathbb{X}, d)$. There then exists a single point $x \in \mathbb{X}$ called a fixed-point of T, verifying $T(x) = x$.

EXAMPLE.–

– The function $f(x) = \frac{1}{2}x + \frac{1}{2}$ ($x \in \mathbb{R}$) has as unique fixed point $x^F = 1$: $f(1) = 1$.

– The function $f(x, y) = (\frac{1}{3}x, \frac{1}{3}y)$ (($x, y) \in \mathbb{R}^2$) has as unique fixed point $x^F = (0, 0) : f(0, 0) = (0, 0)$.

If a function is not contractive, it may have several fixed points, or none.

– The function $f(x) = x^2$ ($x \in \mathbb{R}$) has two fixed points $x_0^F = 0$ and $x_1^F = 1$ ($f(0) = 0$ and $f(1) = 1$).

– The identity function possesses an infinite number of fixed points.

– The function $f(x) = x + 1$ ($x \in \mathbb{R}$) has no fixed point.

1.1.2. *Self-similarity*

DEFINITION.– *Let $(\mathbb{X}, d)$ be a complete metric space and $(\mathcal{H}(\mathbb{X}), d_\mathbb{X})$ the space of objects associated with it. $K \in \mathcal{H}(\mathbb{X})$ is self-similar if there exists a finite number N of contractive transformations on $\mathbb{X}$, $T_i, i = 0 \cdots, N - 1$, such that:*

$$K = \bigcup_{i=0}^{N-1} T_i(K)$$

We will then say that the object K possesses the self-similarity property relating to the transformation set $\{T_i, i = 0 \cdots, N - 1\}$. It is useful to identify transformations using indices to differentiate them and to then explain formulations and algorithms. The order, however, is arbitrary and has no impact on the geometry of objects. We denote by $\Sigma = \{0, \cdots, N - 1\}$, the set of these indices.

The previous definition specifies the intuitive idea introduced in section 1.1. The objects that we are looking at are compacts subsets of a metric space, that is, closed and bounded sets. The nature of the transformations describing self-similarity expresses what we mean by "composed of copies of itself". These transformations are supposed to be contractive, which seems natural because part of the object must be smaller than the object itself. This condition of contraction is sufficient but not necessary. A part is necessarily included in the object, but locally it could be stretched over certain areas, or in a given direction, and contracted in another direction (see Figure 1.5).

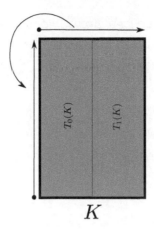

Figure 1.5. *Example of self-similarity involving non-contractive transformations, K = T₀(K) ∪ T₁(K). For a color version of this figure, see www.iste.co.uk/gentil/geometric.zip*

COMMENTS ON FIGURE 1.5.– *The transformation T_0 transforms the overall rectangle K (whose edge is a thick black line) over the blue rectangle by rotating and contracting in one direction and dilating in the other: the edge referred to by the horizontal arrow is transformed into the edge referred to by the vertical edge. The transformation T_1 operates analogously for the pink rectangle.*

1.1.3. *Examples*

1.1.3.1. *The Cantor set*

One of the simplest examples is that of the Cantor set. It is iteratively defined from the $[0, 1]$ segment. This segment is divided into three segments of equal length and the middle segment is removed. The same process is applied to each of the remaining segments (see Figure 1.6). The operation is repeated endlessly. The transformations of $\mathbb{R}$ describing this self-similarity are: $T_0(x) = \frac{x}{3}$ and $T_1(x) = \frac{x}{3} + \frac{2}{3}$.

Figure 1.6. *The Cantor set successively represented at the iteration levels from 1 to 5. For a color version of this figure, see www.iste.co.uk/gentil/geometric.zip*

1.1.3.2. *The Cartesian product of two Cantor sets*

The Cartesian product of two Cantor sets can be directly obtained from the definition of the Cartesian product, or by simply defining self-similarity, as presented by Figure 1.7. This self-similarity can be described by a set of four transformations of $\mathbb{R}^2$:

$$T_0 \begin{pmatrix} x \\ y \end{pmatrix} = \begin{pmatrix} \frac{1}{3} & 0 \\ 0 & \frac{1}{3} \end{pmatrix} \begin{pmatrix} x \\ y \end{pmatrix}, \quad T_1 \begin{pmatrix} x \\ y \end{pmatrix} = \begin{pmatrix} \frac{1}{3} & 0 \\ 0 & \frac{1}{3} \end{pmatrix} \begin{pmatrix} x \\ y \end{pmatrix} + \begin{pmatrix} \frac{2}{3} \\ 0 \end{pmatrix}$$

$$T_2 \begin{pmatrix} x \\ y \end{pmatrix} = \begin{pmatrix} \frac{1}{3} & 0 \\ 0 & \frac{1}{3} \end{pmatrix} \begin{pmatrix} x \\ y \end{pmatrix} + \begin{pmatrix} \frac{2}{3} \\ \frac{2}{3} \end{pmatrix}, T_3 \begin{pmatrix} x \\ y \end{pmatrix} = \begin{pmatrix} \frac{1}{3} & 0 \\ 0 & \frac{1}{3} \end{pmatrix} \begin{pmatrix} x \\ y \end{pmatrix} + \begin{pmatrix} 0 \\ \frac{2}{3} \end{pmatrix}$$

Figure 1.7. *Cartesian product of two Cantor sets successively represented at iteration levels from 1 to 4. For a color version of this figure, see www.iste.co.uk/gentil/geometric.zip*

1.1.3.3. *The Sierpinski triangle*

The Sierpinski triangle is also a classic example. It is built from a solid triangle, subdivided, and four triangles, from which the central triangle is removed. The operation is iterated for each of the resulting triangles (see Figure 1.8). The three transformations of $\mathbb{R}^2$ for this Sierpinski triangle are homotheties of ratio $\frac{1}{2}$ centered at a vertex of the triangle:

$$T_0 \begin{pmatrix} x \\ y \end{pmatrix} = \begin{pmatrix} \frac{1}{2} & 0 \\ 0 & \frac{1}{2} \end{pmatrix} \begin{pmatrix} x \\ y \end{pmatrix} - \begin{pmatrix} \frac{1}{2} \\ 0 \end{pmatrix}$$

$$T_1 \begin{pmatrix} x \\ y \end{pmatrix} = \begin{pmatrix} \frac{1}{2} & 0 \\ 0 & \frac{1}{2} \end{pmatrix} \begin{pmatrix} x \\ y \end{pmatrix} + \begin{pmatrix} \frac{1}{2} \\ 0 \end{pmatrix}$$

$$T_2 \begin{pmatrix} x \\ y \end{pmatrix} = \begin{pmatrix} \frac{1}{2} & 0 \\ 0 & \frac{1}{2} \end{pmatrix} \begin{pmatrix} x \\ y \end{pmatrix} + \begin{pmatrix} 0 \\ \frac{1}{2} \end{pmatrix}$$

1.1.3.4. *The Menger sponge*

The Menger sponge is built from a square decomposed into nine squares of the same size, removing the central square. Figure 1.9 illustrates the recursive construction. Self-similarity is described by eight homotheties of ratio $\frac{1}{3}$, with the respective centers of the four vertices of the square and the four middles of the sides square.

Figure 1.8. *The Sierpinski triangle successively represented at iteration levels from 1 to 4. For a color version of this figure, see www.iste.co.uk/gentil/geometric.zip*

Figure 1.9. *The Menger sponge successively represented at iteration levels from 1 to 4. For a color version of this figure, see www.iste.co.uk/gentil/geometric.zip*

1.1.3.5. *The Romanesco broccoli*

The Romanesco broccoli is a very nice illustration of a self-similar structure produced by nature. It is composed of a multitude of growths, each resembling a miniature copy of the whole broccoli. By using this self-similar property, it is possible to simply generate its geometric structure using seven transformations of $\mathbb{R}^3$ (see Figure 1.10). The transformations describing the self-similarity of the figure are given as an appendix in section A.1.1.

1.1.3.6. *Non-fractal self-similar object*

In previous examples, we very clearly perceive the fractal aspect of structures, which is reflected in increasing numbers of ramifications, or details that are always structured in the same way. However, not every self-similar object is a fractal. A very simple example is the segment that can be decomposed into two smaller segments (see Figure 1.11). There are infinite possibilities to describe this self-similarity using two transformations of $\mathbb{R}$: $T_0^\alpha(x) = \alpha x$, $T_1^\alpha(x) = (1 - \alpha)x + \alpha$ with $\alpha \in]0, 1[$. In fact, regardless of the value of $\alpha \in]0, 1[$, $T_0^\alpha([0, 1]) \cup T_1^\alpha([0, 1]) = [0, 1]$. An arbitrary number of transformations can also be used. Similarly, the square can be broken down into smaller squares. As with the segment, this decomposition is not unique.

1.1.3.7. *L-shape*

Other less obvious forms have properties of self-similarity as an L-shape (see Figure 1.12).

Figure 1.10. *Example of Romanesco broccoli consisting of seven self-similar elements. For a color version of this figure, see www.iste.co.uk/gentil/geometric.zip*

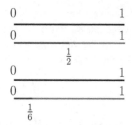
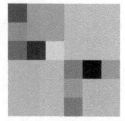

Figure 1.11. *On the left-hand side, we provide a few examples of self-similarity describing the segment. At the top, the segment is composed of two segments of equal length $\alpha = \frac{1}{2}$. At the bottom, it is cut into two segments, the first of length $\frac{1}{6}$ and the second of length $\frac{5}{6}$ ($\alpha = \frac{1}{6}$). On the right side, we provide an example of self-similarity for a square. For a color version of this figure, see www.iste.co.uk/gentil/geometric.zip*

REMARK.– There is no unicity in the description of the self-similarity property. If we have:

$$K = \bigcup_{i \in \Sigma} T_i(K)$$

by replacing K in the second member of the equality with its self-similar decomposition, we get:

$$K = \bigcup_{i,j \in \Sigma} T_i \circ T_j(K)$$

The set of transformations defined by the compositions of all the pairs of initial transformations, namely $\{T_i \circ T_j\}_{i,j \in \Sigma}$, is another way of seeing or expressing this same self-similarity. Another example is the description of the self-similarity of the segment $[0, 1]$ presented above.

Figure 1.12. *Example of a decomposition of an L-shape into several similar elements. For a color version of this figure, see www.iste.co.uk/gentil/geometric.zip*

The question here is whether this self-similarity property is relevant to be used as a model of geometric representation. In other words, can the mere knowledge of this property be sufficient in defining, describing and manipulating an object? In geometric modeling, it is useful that a representation verifies a number of properties such as validity, non-ambiguity, unicity, brevity, ease of construction and adequacy of applications (see (Requicha 1996)). Without these properties, this representation will be more complex to use. For example, if the representation can lead to non-valid descriptions (that is, not representing a coherent object), it will be necessary to develop devices to verify the validity of a given representation. In addition, certain automatic operations or processes applied to valid representations may produce non-valid, non-manufacturable and unusable representations. We know that for a given self-similar object, there is no unicity in the representation by a set of contractive transformations. This is not without a number of problems, for example in searching for identical objects in a database. This is nonetheless the case for most existing representations: meshes, CSG model, B-rep, etc.

From this point of view, the following property is fundamental because alone, it justifies the relevance of self-similarity and its use as a system of representation.

PROPERTY.– Let $\{T_i\}_{i \in \Sigma}$ be a set of N transformations on $\mathbb{X}$, then there exists a unique element $K \in \mathcal{H}(\mathbb{X})$, possessing the self-similarity property relating to the contractive transformations $T_i, i = 0 \cdots, N - 1$, that is, verifying:

$$K = \bigcup_{i \in \Sigma} T_i(K)$$

The proof is given by Hutchinson (1981). It is based on the contractive property of an operator called the Hutchinson operator.

DEFINITION.– *In the early 80s, Hutchinson (1981) used the Banach fixed point theorem to deduce the existence and the uniqueness of an attractor for a hyperbolic IFS, i.e. the fixed point of the associated contractive map. Let $\{T_i\}_{i \in \Sigma}$ be a set of N contractive transformation on $\mathbb{X}$, and we call the Hutchinson operator associated with $\{T_i\}_{i \in \Sigma}$, the operator $\mathbb{T} : \mathcal{H}(\mathbb{X}) \to \mathcal{H}(\mathbb{X})$ defined by:*

$$\forall K \in: \mathcal{H}(\mathbb{X}), \quad \mathbb{T}(K) = \bigcup_{i \in \Sigma} T_i(K)$$

PROPERTY.– Let $\{T_i\}_{i \in \Sigma}$ be a set of N contractive transformations on $\mathbb{X}$, then the associated Hutchinson operator $\mathbb{T}$ is contractive in $(\mathcal{H}(\mathbb{X}), d_{\mathbb{X}})$.

PROOF.– See Barnsley (1988).

The proof of the existence and unicity of an object, verifying the self-similarity property described by a set of contractions, is then simple.

PROOF.– The Hutchinson operator $\mathbb{T}$ is contractive in the complete metric space $(\mathcal{H}(\mathbb{X}), d_{\mathbb{X}})$. According to the Banach fixed-point theorem, there exists a unique element $K \in \mathcal{H}(\mathbb{X})$ verifying:

$$\mathbb{T}(K) = K$$

By definition of $\mathbb{T}$, K has the self-similarity property relatively to $\{T_i\}_{i \in \Sigma}$:

$$K = \bigcup_{i \in \Sigma} T_i(K) \qquad \qquad \Box$$

Therefore, a set of contractive transformations defines a unique object. This object has the self-similarity property described by this set of transformations. From a geometric modeling point of view, this result has an important consequence. For an object $K \in \mathcal{H}(\mathbb{X})$, if we can exhibit a set of contractive transformations $\{T_i\}_{i \in \Sigma}$ such that $K = \bigcup_{i=0}^{N-1} T_i(K)$, by unicity of the fixed point, the set K is determined without ambiguity by this set of transformations $\{T_i\}_{i \in \Sigma}$.

Thereby, the object in Figure 1.13 presents an object of "complex" structure, whose description by means of a standard model would be difficult, if not impossible. Using the self-similarity property, this object is completely determined by two affine transformations T_0 and T_1.

Another issue is to determine whether an object is self-similar, and to find the set of transformations corresponding to this self-similarity. In this book, we will not directly address this issue, which is known as the inverse problem. Our goal is to provide tools to design and manipulate self-similar objects. The inverse problem is a difficult problem and is only partially solved. Solutions have been proposed for a few specific cases (Barnsley *et al.* 1986; Collet *et al.* 2000; Guérin and Tosan 2005).

Figure 1.13. *Example of self-similarity. The object on the left-hand side has a self-similarity highlighted by the right-hand side illustration. The blue portion is a copy using a homothety while the pink portion is obtained by applying the composition of a homothety with a symmetry, inverting the figure following an axis. For a color version of this figure, see www.iste.co.uk/gentil/geometric.zip*

1.1.4. *Iterated function systems*

In this section, we give the basic definitions and properties of iterated function systems, which will be subsequently useful.

A large amount of research focuses on the properties of iterated function systems and their attractors. The purpose of this book is not to survey them. We emphasize a few properties that we believe are essential for the construction of the BC-IFS model that we shall introduce later in this section.

DEFINITION.– *Let $(\mathbb{X}, d)$ be a metric space, an iterated function system (denoted IFS) is a finite set of contractive transformations $\mathbb{T} = \{T_i\}_{i=0}^{N-1}$ on $\mathbb{X}$. We denote by $\Sigma = \{0, \ldots, N-1\}$ the finite set of the transformation indices $|\Sigma| = N$. The IFS is denoted by $\{\mathbb{X}; T_i \mid i \in \Sigma\}$ or $\{T_i \mid i \in \Sigma\}$ if there is no ambiguity about the space $\mathbb{X}$.*

REMARK.– In order to simplify the notation, the IFS is denoted in the same way as the associated Hutchinson operator. These are nonetheless two different mathematical entities. However, there can be no ambiguity within the context.

DEFINITION.– *Let $\{T_i\}_{i\in\Sigma}$ be un IFS and $\mathbb{T}$ the associated Hutchinson operator. The only fixed point of $\mathbb{T}$ is called an attractor and is denoted $\mathcal{A}(\mathbb{T})$, or simply $\mathcal{A}$ if there is no ambiguity.*

REMARK.– If $\mathbb{T} = \{T\}$ is reduced to a transformation, then $\mathcal{A}(\mathbb{T}) = \{x^F\}$ where $x^F \in \mathbb{X}$ refers to the fixed point of the transformation T.

PROPERTY.– Let $\mathbb{T}^0$ and $\mathbb{T}^1$ denote two IFS on $\mathbb{X}$ such that $\mathbb{T}^0 \subset \mathbb{T}^1$, then:

$$\mathcal{A}(\mathbb{T}^0) \subset \mathcal{A}(\mathbb{T}^1)$$

The lattice structure of the sets is transported to that of the attractors by the function $\mathcal{A}$, which associates its attractor with an IFS. An example of this lattice structure in attractors is given in Figure 1.14. This property is essential: it induces a B-rep structure on the attractors and we shall make use of it to control the topological structure of fractal objects. Figure 1.15 shows how this property can be utilized to connect two attractors with a common "sub-attractor".

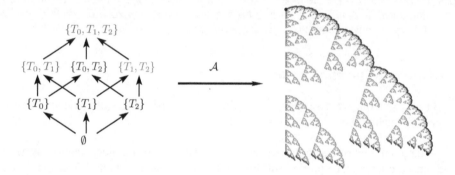

Figure 1.14. *Lattice structure of the attractors. On the left, the lattice structure on the set of the parts of an IFS is transported by $\mathcal{A}$ over the set of the associated attractors (right-hand side). For a color version of this figure, see www.iste.co.uk/gentil/geometric.zip*

COMMENTS ON FIGURE 1.14.– *The fixed points of each transformation T_0, T_1 and T_2 represented by black dots are included in the attractors $\mathcal{A}(\{T_i, T_j\}), i \neq j$, which are themselves included in $\mathcal{A}(\{T_0, T_1, T_2\})$. For example, the attractor of the sub-IFS $\{T_1, T_2\}$, represented in blue, is a subset of the global attractor.*

PROPERTY.– Let $\mathbb{T} = \{T_i\}_{i \in \Sigma}$ be an IFS, $\mathcal{A}(\mathbb{T}) \supseteq \{x_i^F\}_{i \in \Sigma}$, where x_i^F refers to the fixed point of the transformation T_i.

PROOF.– This property is a consequence of the previous property. $\forall i \in \Sigma, \{T_i\} \subset \mathbb{T}$, then $\mathcal{A}(\{T_i\}) = \{x_i^F\} \subset \mathcal{A}(\mathbb{T})$.

PROPERTY.– Let $\mathbb{T} = \{T_i\}_{i \in \Sigma}$ be an IFS; if $x \in \mathcal{A}(\mathbb{T})$, then $\forall i \in \Sigma, T_i(x) \in \mathcal{A}(\mathbb{T})$.

PROOF.– $\mathcal{A}(\mathbb{T}) = \mathbb{T}(\mathcal{A}(\mathbb{T})) = \bigcup_{i \in \Sigma} T_i(\mathcal{A}(\mathbb{T}))$.

Determining the shape of an attractor from the knowledge of an IFS is a difficult exercise, even in the case of simple affine transformations of the Euclidian plane.

Figure 1.15. *An example of a connection between two attractors. The green attractor was designed from the transformations* $\{T_1, T_2\}$ *originating from the IFS* $\{T_0, T_1, T_2\}$ *in Figure 1.14 and by adding an additional transformation* T_3. *Therefore, the attractors* $\mathcal{A}(\{T_0, T_1, T_2\})$ *(in pink) and* $\mathcal{A}(\{T_1, T_2, T_3\})$ *(in green) have a common sub-attractor,* $\mathcal{A}(\{T_1, T_2\})$ *(in blue), depending on the one that they connect to. For a color version of this figure, see www.iste.co.uk/gentil/geometric.zip*

REMARK.– In Barnsley (1988), Barnsley distinguishes between transformation sets, called IFS, and contractive transformation sets, called hyperbolic IFS. In this book, we assume that the transformations of an IFS are contractive. It is possible to build IFS composed of non-contractive transformations with an attractor, but this goes beyond the scope of this book (Collet *et al.* 2000; Barnsley and Vince 2013a).

PROPERTY.– Let $\mathbb{T} = \{T_i\}_{i\in\Sigma}$ be an IFS and T an invertible transformation $\mathbb{X}$, such that $\forall i \in \Sigma$, TT_iT^{-1} is contractive. The transform of $\mathcal{A}(\mathbb{T})$ by T is given by the IFS $\{TT_iT^{-1}\}_{i\in\Sigma}$.

PROOF.–

$$\bigcup_{i\in\Sigma} TT_iT^{-1}T(\mathcal{A}) = \bigcup_{i\in\Sigma} TT_i(\mathcal{A})$$

$$= T\bigcup_{i\in\Sigma} T_i(\mathcal{A})$$

$$= T(\mathcal{A})$$

Due to the unicity of the fixed point, $T(A)$ is the attractor of $\{TT_iT^{-1}\}_{i\in\Sigma}$.

PROPERTY.– Let $\mathbb{T} = \{T_i\}_{i\in\Sigma}$ be an IFS and M an involution $\mathbb{X}$ (that is, $M^2 = Id$). Then, the attractor associated with the IFS $\mathbb{T}'\{T_i, MT_iM\}_{i\in\Sigma}$ is invariant by M and:

$$\mathcal{A}(\mathbb{T}) \subset \mathcal{A}(\mathbb{T})$$
$$\mathcal{A}(M(\mathbb{T})) \subset \mathcal{A}(\mathbb{T}')$$

where $M(\mathbb{T})$ designates $\mathbb{T}\{T_i, MT_iM\}_{i\in\Sigma}$.

PROOF.– Since M is an involution $M = M^{-1}$, then:

$$
\begin{aligned}
M(\mathcal{A}(\mathbb{T}')) &= \mathcal{A}(\{MT_iM, MMT_iMM\}_{i\in\Sigma}) \\
&= \mathcal{A}(\{MT_iM, T_i\}_{i\in\Sigma}) \\
&= \mathcal{A}(\mathbb{T}')
\end{aligned}
$$

1.1.5. *Visualization and approximation*

First of all, we shall look at one of the most basic evaluations of a representation of a geometric object: visualization.

How do you visualize the attractor of an IFS by only using the data of the set of transformations describing the self-similarity of the fractal?

For this purpose, we simply make use of a corollary of the fixed-point theorem.

PROPERTY.– Let T be a contractive transformation in a complete metric space $(\mathbb{X}, d)$, then $\forall x_0 \in \mathbb{X}$:

$$
\lim_{n\to\infty} T^n(x_0) = x^F
$$

where x^F represents the fixed point of T and $T^n = \underbrace{T \circ \cdots \circ T}_{n\,times}$.

By applying this property to the Hutchinson operator, we deduce an algorithm for constructing a sequence of objects $\{K_n\}_{n\in\mathbb{N}}$, whose limit is the attractor of the IFS. One simply has to choose an arbitrary compact set K of $\mathbb{X}$, a number of iterations n depending on the desired approximation (we shall discuss this point later) and calculate $K_n = \mathbb{T}^n(K)$. For $\mathbb{T} = \{T_0, T_1\}$, this yields:

$$
\begin{aligned}
K_0 &= K \\
K_1 &= \mathbb{T}(K_0) = T_0(K) \cup T_1(K) \\
K_2 &= \mathbb{T}(K_1) = T_0(T_0(K_0) \cup T_1(K)) \cup T_1(T_0(K) \cup T_1(K)) \\
&\vdots \qquad \vdots \qquad\quad \vdots
\end{aligned}
$$

The computation of the transformation of the primitive $T(K_0)$ can be more or less complex depending on the nature of the transformation and of the representation chosen to describe the object. However, regardless of the transformation T and a set

A, the transform of A by T is defined by $T(A) = \{T(x)|x \in A\}$. Therefore, we notice that the transform of the union of the sets A and B is:

$$
\begin{aligned}
T(A \cup B) &= \{T(x)|x \in A \cup B\} \\
&= \{T(x)|x \in A\} \cup \{T(x)|x \in B\} \\
&= T(A) \cup T(B)
\end{aligned}
$$

The calculation of K_n can then be presented as follows:

$$
\begin{aligned}
K_0 &= K \\
K_1 &= \mathbb{T}(K_0) & &= T_0(K) \cup T_1(K) \\
K_2 &= \mathbb{T}(K_1) & &= T_0 T_0(K) \cup T_0 T_1(K) \cup T_1 T_0(K) \cup T_1 T_1(K) \\
K_3 &= \mathbb{T}(K_2) & &= T_0 T_0 T_0(K_0) \cup T_0 T_0 T_1(K) \cup T_0 T_1 T_0(K_0) \cup T_0 T_1 T_1(K_0) \\
& & & \quad \cup T_1 T_0 T_0(K) \cup T_1 T_0 T_1(K) \cup T_1 T_1 T_0(K) \cup T_1 T_1 T_1(K)
\end{aligned}
$$

$$
\vdots \qquad \vdots \qquad\qquad \vdots
$$

$$
K_n = \mathbb{T}(K_{n-1}) = \bigcup_{\alpha_i \in \{0,1\}} T_{\alpha_1} \ldots T_{\alpha_n}(K)
$$

This calculation amounts to evaluating all possible compositions of transformations of n, among the N transformations of the IFS. It remains to then transform the primitive K_0 by each of these compositions of transformations and to make a union of the results. We can represent this computation by what we call *the evaluation tree* of the attractor (see Figure 1.16).

1.1.6. *Computer implementation*

Several algorithms can be proposed for computing K_n. K_n is a union of compositions of transformations applied to K. With the union operator being commutative ($A \cup B = B \cup A$), we can evaluate the compositions of the transformations in any order.

The evaluation tree proposes an implicit order due to the fact that the composition with the transformations is done by composition on the right-hand side. A composition on the left side gives the same end result. However, with the right-hand side composition, we have the following property: by denoting T_{σ_l} the composition of the transformations obtained at a node of depth l, $T_{\sigma_l}(\mathcal{A}) \subset T_{\sigma_l} \circ T_i(\mathcal{A}), \forall i \in \Sigma$. That is, each part of the attractor corresponding to a node contains the parts of the attractor associated with the sub-nodes. This property is interesting for optimizing given processes, such as trimming the shape for displaying or performing a computation of the intersection with another object: if there is no intersection at the node level, it is not necessary to evaluate the sub-nodes. Unfortunately, this property is not directly exploitable since the attractor $\mathcal{A}$ is not known, or even computable.

Nonetheless, it is possible to utilize the following property to determine a compact set K, from which the inclusion property of the nodes of the evaluation tree at their peer node is verified.

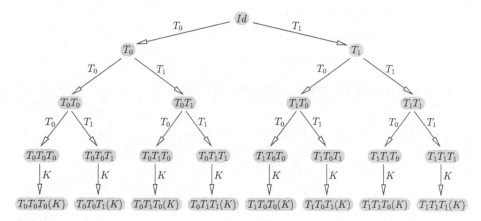

Figure 1.16. *The evaluation tree of the attractor of the IFS computed at the third level. The internal nodes correspond to the computation of the compositions of the transformations. The union of the leaves corresponds to K_3*

PROPERTY.– Let $\mathbb{T}$ be an IFS and $K \in \mathcal{H}(\mathbb{X})$ such that $\mathbb{T}(K) \subseteq K$, then the evaluation tree is such that each node contains its sub-nodes. A special case is $K = \mathcal{A}$.

A detailed description of the method for finding such a K is given in Mishkinis *et al.* (2012).

For these reasons, it is the right composition algorithm that we choose to present.

For each branch of the evaluation tree, we must calculate $T_{\sigma_0} \circ T_{\sigma_1} \circ \cdots \circ T_{\sigma_n}(K)$, where $\sigma_0\sigma_1 \cdots \sigma_n$ represents the path in the tree. For this expression, we can compute the composition of the transformations $T = T_{\sigma_0} \circ T_{\sigma_1} \circ \cdots \circ T_{\sigma_n}$ and then apply it to the primitive: $T(K)$. This computation corresponds to Algorithm 1.1a. Another solution consists of applying the rightmost transformation to the primitive $T_{\sigma_n}(K)$, then applying the following transformation, in the right-to-left order: $T_{\sigma_{n-1}}(T_{\sigma_n}(K))$ and so on. In the latter case, we get Algorithm 1.1b. For these first two algorithms, the smallest amount of memory is used because we evaluate the tree in a branch-wise manner (in-depth exploration). The elemental parts of the attractor (namely, $T_\sigma(\mathcal{A}) \subset T_\sigma \circ T_i(\mathcal{A})$) are evaluated one after the other and as soon as they are calculated, they are displayed. However, at no time do we have a complete data structure that represents the approximation of the $\mathbb{T}^n(K)$ attractor. In this case, the display device acts as a support for the union of the parts. The construction of the union could be integrated into the algorithms.

The choice of either one can be guided by different criteria. A determining factor is knowing whether we are able to calculate the composition of two transformations effectively. This computation may be impossible or costly if the transformations are transcendental, polynomial or piece-wise functions. In this case, we shall use Algorithm 1.1b. We did not specify how to actually calculate $T_i(K)$. In practice, the transformations that we use are affine or linear transformations. This has the advantage of being able to represent them in matrix form. In this case, both algorithms can be used, by just changing the composition of the transformations $T \circ T_i$ by the matrix product $M_T \times M_{T_i}$.

Finally, one criterion that can be used in choosing the algorithm is the number of operations, that is, the complexity of the algorithms. Consider the case of affine transformations, with a representation in matrix form $(p \times p)$ in homogeneous coordinates and a primitive represented in the form of a list of points. If the primitive consists of a single point, the calculation of $T_i(K)$ requires $(2p - 1)$ operations. For a tree depth of n, this represents $2np + n$ operations. If we compute the product of n $(p \times p)$ matrices that we apply to the point, we must achieve $2np^2 - (n - 2)p - 1$ operations.

REMARK.– The proposed algorithms use the fact that $T(A \cup B) = T(A) \cup T(B)$. To calculate K_n, this property does not necessarily simplify the calculation. This again depends on the representation of the objects and the computation complexity of the transformation of a union, and that of the union of the transformations. Concerning the visualization, the gain is clear because the display algorithms are responsible for joining the transformations of the primitive K.

To apply a transformation (affine or linear) to the primitive K, a very simple solution consists of representing the primitive by a mesh. Then for each point of the mesh, one will simply have to compute the product of the matrix representing the transformation by the column vector composed of the coordinates of the point. These points can also be stored in a matrix:

$$P = \begin{pmatrix} P_{0_x} & P_{1_x} & \cdots & P_{l_x} \\ P_{0_y} & P_{1_y} & \cdots & P_{l_y} \\ P_{0_z} & P_{1_z} & \cdots & P_{l_z} \end{pmatrix}$$

The evaluation then consists of performing matrix products only:

$$K_n = \bigcup_{\alpha_i \in \Sigma} M_{T_{\alpha_1}} \times M_{T_{\alpha_n}} \times P$$

REMARK.– We can proceed similarly with a primitive represented by NURBS. It suffices to transform the control points, Nonetheless, it is often unnecessary to use sophisticated representations because the calculation of the approximation of the attractor converges toward the same result, regardless of the primitive chosen.

Ensure: $K_n = \mathbb{T}(K_0)$
Require: $\{T_i\}_{i \in \Sigma}, n > 0, K \in \mathcal{H}(\mathbb{X})$
1: $T \leftarrow Id_{\mathbb{X}}$ ▷ Identity transformation of $\mathbb{X}$
2: DisplayAttractor(T,n)
3: **procedure** DISPLAYATTRACTOR(T,m)
4:      **if** $m = 0$ **then**
5:          $K_{tmp} \leftarrow T(K)$
6:          Print K_{tmp}
7:      **else**
8:          **for** $i \in \Sigma$ **do**
9:              DisplayAttractor($T \circ T_i$,$m - 1$)
10:         **end for**
11:     **end if**
12: **end procedure**

---

a)

Ensure: $K_n = \mathbb{T}(K_0)$
Require: $\{T_i\}_{i \in \Sigma}, n > 0, K \in \mathcal{H}(\mathbb{X})$
    $T \leftarrow Id_{\mathbb{X}}$ ▷ Identity transformation of $\mathbb{X}$
2: DisplayAttractor(K_0,n)
    function DISPLAYATTRACTOR(K,m)
Require: $K \in \mathcal{H}(\mathbb{X})$ ▷ transformed primitive
4:      **if** $m = 0$ **then**
            Print(K)
6:      **else**
            for $i \in \Sigma$ **do**
8:              DisplayAttractor($T_i(K)$,$m - 1$)
            end for
10:     **end if**
        end function

---

b)

Algorithm 1.1. *Visualization K_n*

1.1.7. *Notions of address and parameterization*

In geometric modeling, the parameterization of shapes is a very useful tool: to apply a texture, to analyze and control differential properties, to visualize the shape, etc. This involves providing a parameter space and determining a function that associates each parameter value with a point. This parameterization function should verify a number of properties such as surjectivity and continuity, at least C_0. For curves and surfaces, which are "standard" topological structures, it is obvious to choose the parameterization domain: classically $[0,1]$ for a curve, or $[0,1]^2$ for a surface. However, for the IFS attractors, the choice of parameter space is less obvious because their topological structures can be "non-standard" and very varied (set of accumulation points, wired or porous structure, etc.). Nonetheless, it is possible to define a parameterization that can be used regardless of the attractor. This parameterization is based on the notion of address. The intuitive idea is very simple. Consider a path in the evaluation tree going from the root to a leaf. A node of depth j represents the composition of the transformations $T_{\sigma_0} \circ \cdots \circ T_{\sigma_j}$. We can associate it with the sub-part of the attractor $T_{\sigma_0} \circ \cdots \circ T_{\sigma_j}(\mathcal{A})$. Therefore, the finite sequence of indices $(\sigma_0 \cdots \sigma_j)$ represents this sub-part of the attractor. At an additional iteration, $T_{\sigma_0} \circ \cdots \circ T_{\sigma_j} \circ T_{\sigma_{j+1}}$ corresponds to a part of $T_{\sigma_0} \circ \cdots \circ T_{\sigma_j} \circ T_{\sigma_{j+1}}(\mathcal{A})$, representing a smaller area (given that transformations are contractive). Proceeding forward, we understand that $T_{\sigma_0} \circ \cdots \circ T_{\sigma_n}(\mathcal{A})$ tends toward an increasingly small area of the attractor when n increases. By tending n towards infinity, the expression $T_{\sigma_0} \circ \cdots \circ T_{\sigma_n}(\mathcal{A})$ tends toward a point of the attractor. Thereby, each point of the attractor will be represented by an infinite word composed of indices belonging to Σ. Let us formally explain the notions of address and the addressing function (Barnsley 1988).

THEOREM.– Let $\mathbb{T} = \{T_i\}_{i \in \Sigma}$ be an IFS indexed by a set $\Sigma = \{0, \cdots, N-1\}$ defined over a metric space $(\mathbb{X}, d)$. Let p be a point of $\mathbb{X}$. Let $\sigma = \{\sigma_0 \cdots \sigma_n \cdots\}$ be an infinite word of Σ. Then:

$$\lim_{n \to \infty} T_{\sigma_0} \circ \cdots \circ T_{\sigma_n}(p)$$

exists and belongs to $\mathcal{A}(\mathbb{T})$ and is independent of p.

The function ϕ defined on the set Σ^ω of the words of Σ by:

$$\phi(\sigma) = \lim_{n \to \infty} T_{\sigma_0} \circ \cdots \circ T_{\sigma_n}(p)$$

is continuous and surjective. If $p \in K$ is a compact subset of $\mathbb{X}$, then convergence is uniform. The function ϕ is called the address function.

PROOF.– See (Barnsley 1988).

This theorem justifies the parameterization of the points of the attractor by the set of infinite words of Σ. This type of parameterization is unusual for a geometric shape. This may seem strange in view of the standard parameterizations mentioned above and not easy to manipulate. However, there are many advantages to this approach from the point of view of computer processing:

– the definition of the address function provides an immediate algorithm to evaluate the point of the corresponding attractor. The transformations just have to be composed according to the indices composing the infinite word;

– this is equivalent to evaluating a branch of the evaluation tree that would be built indefinitely;

– obviously, we have to set a finite number of iterations (the depth of the tree) to actually perform the computations. Then, we only obtain an approximation of the value of $\phi(\sigma)$, but this approximation can be as accurate as we want.

In order to ease the notation of infinite words, we shall use the standard notion where $i^\omega = i \cdots i \cdots$ represents an infinity of repetitions of index i. Similarly, if $\sigma_0 \cdots \sigma_n$ is a finite word, $(\sigma_0 \cdots \sigma_n)^\omega$ represents an infinity of concatenations of the word $\sigma_0 \cdots \sigma_n$.

REMARK.– Let us consider the IFS $\mathbb{T} = \{\mathbb{R}; T_0(x) = \frac{x}{2}, T_1(x) = \frac{x}{2} + \frac{1}{2}\}$. Its attractor is $\mathcal{A}(\mathbb{T}) = [0, 1]$. All the points of $T_0(\mathcal{A}) = [0, \frac{1}{2}]$ have an address starting with 0 and all the points of $T_1(\mathcal{A}) = [\frac{1}{2}, 1]$ have an address starting with 1. Similarly, the points:

– of $T_0 \circ T_0(\mathcal{A}) = [0, \frac{1}{4}]$ have an address starting with 00;

– of $T_0 \circ T_1(\mathcal{A}) = [\frac{1}{4}, \frac{1}{2}]$ have an address that starts with 01;

– of $T_1 \circ T_0(\mathcal{A}) = [\frac{1}{2}, \frac{3}{4}]$ have an address that starts with 10;

– of $T_1 \circ T_1(\mathcal{A}) = [\frac{3}{4}, 1]$ have an address that starts with 11.

Continuing with the reasoning, we notice that the address of a point $x \in \mathcal{A} = [0, 1]$ corresponds to its dyadic decomposition:

$$x = \sum_{i=0}^{\infty} \frac{\sigma_i}{2^{(i+1)}} \Leftrightarrow x = \phi(\sigma_0 \sigma_1 \cdots \sigma_n \cdots)$$

The coding of a real number $x \in [0, 1]$ in decimal base is equivalent to its address for the IFS decomposing the interval $[0, 1]$ in ten-part lengths equal to one-tenth: $\mathbb{T} = \{T_i(x) = \frac{x}{10} + \frac{i}{10}\}, i = \{0, 1, \cdots, 9\}$

As for real numbers, there may be several possible codings (for example, $0.5 = 4.99999$). For these points, we shall also have several possible addresses.

These multiple address points appear as soon as there are two sub-parts of the attractor intersecting one another, that is $\exists i, j \in \Sigma \; tq \; T_i(\mathcal{A}) \cap T_j(\mathcal{A}) \neq \emptyset$.

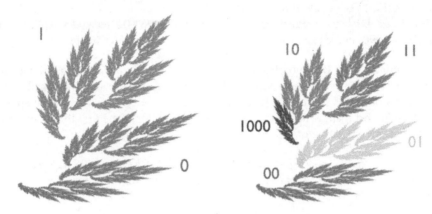

Figure 1.17. *Example of the parameterization of the attractor in Figure 1.13. On the right side, all image points of the attractor by the transformation T_0 have an address starting with 0. The image points of the attractor by T_1 have an address that starts with 1, and so on. The image points of the attractor by $T_0 \circ T_1$ have an address starting with 01 (indicated in green on the right-hand side image). For a color version of this figure, see www.iste.co.uk/gentil/geometric.zip*

1.1.8. *Finite address*

The notion of finite address is very useful because it allows one to identify a subset of the attractor. By defining equivalence relations between finite addresses, we should be able to control the topology of fractal structures.

DEFINITION.– *For every finite word $\sigma^n \in \Sigma^n$ of length n, we map:*

$$\mathcal{A}_{\sigma^n} = T_{\sigma_0^n} \circ \cdots \circ T_{\sigma_n^n}(\mathcal{A})$$

PROPERTY.– *If σ^n is a prefix of σ^m, then:*

$$\mathcal{A}_{\sigma^n} \supset \mathcal{A}_{\sigma^m}$$

1.1.9. *IFS morphisms*

IFS morphisms have been introduced by Tosan (1999). They make it possible to identify necessary conditions on the IFS to obtain attractors of the same topology.

These conditions are expressed in the form of constraints on the addressing functions. IFS morphisms can be seen as an extension of *fractal interpolation functions* (FIF) (Barnsley 1986) to any fractal topological structures. In particular, they help establish the precise relationship between IFS, Bezier and B-spline (Zair and Tosan 1996) curves and surfaces. The latter is presented in Chapter 3.

The principle of IFS morphisms is to build two IFS: the attractor of the first corresponds to the parameter space of the second and to the figure defined in the modeling space. The addressing functions will allow the construction of a mapping, called a transport map, which at each point of the attractor of the first IFS associates a point of the second attractor (see Figure 1.18).

DEFINITION.– *Let* $\mathbb{T} = \{T_i; i \in \Sigma\}$ *and* $\mathbb{T}' = \{T_i'; i \in \Sigma\}$ *be two IFS indexed for the same set* Σ. *Let* $\mathcal{A}(\mathbb{T}) = \mathcal{A}$ *and* $\mathcal{A}(\mathbb{T}') = \mathcal{A}'$ *their respective attractors. We define the transport mapping H by:*

$$\forall p \in \mathcal{A}, H(T_i p) = T_i' H(p)$$

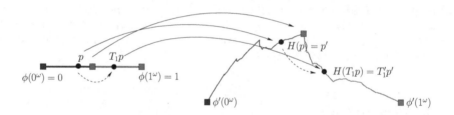

Figure 1.18. *Example of a transport mapping that defines a morphism of IFS. For this example, the morphism defines a curve parameterized by [0,1]. For a color version of this figure, see www.iste.co.uk/gentil/geometric.zip*

PROPERTY.– Let $\mathbb{T} = \{T_i; i \in \Sigma\}$ and $\mathbb{T}' = \{T_i'; i \in \Sigma\}$ be two IFS indexed by the same set Σ and H the associated transport map, then:

$$H(\mathcal{A}) = \mathcal{A}'$$

PROOF.–

$$H(\mathcal{A}) = H(\bigcup_{i \in \Sigma} T_i \mathcal{A})$$

$$= \bigcup_{i \in \Sigma} T_i' H(\mathcal{A})$$

By unicity of the fixed point $H(\mathcal{A}) = \mathcal{A}'$.

The transport maps two attractors. From this property, we deduce that H also transports the entire structure of the attractor, because $\forall i \subset \Sigma, H(T_i(\mathcal{A})) - T_i'(\mathcal{A}')$, that is, the ith subdivision of the attractor $\mathcal{A}$ is transported over the ith subdivision of the attractor $\mathcal{A}'$. And more generally, for any address σ, $H(\phi(\sigma)) = \phi'(\sigma)$, where ϕ and ϕ' represent the address functions of $\mathbb{T}$ and $\mathbb{T}'$, respectively. The map H associates the points of A of address σ with the point of $\mathcal{A}'$ with the same address. Therefore, in order for H to be well defined, it is necessary that the image of any point p with multiple addresses has the same addresses. If the transformations T_i are invertible, this condition is expressed as follows:

$$\forall p \in \mathcal{A} \text{ such that } p \in T_i(\mathcal{A}) \cap T_j(\mathcal{A}) \Rightarrow T_i'(H(T_i^{-1}(p)) = T_j'(H(T_j^{-1}(p)) $$

$$[1.1]$$

The example shown in Figure 1.19 does not define a morphism between the two attractors, since the point $\frac{1}{2}$ (indicated by a green square on the left attractor) has two addresses (10^ω and 01^ω), whereas these two addresses represent distinct points on the right attractor. On the other hand, Figure 1.18 does show an IFS morphism.

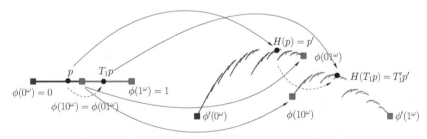

Figure 1.19. *Example of mapping between two attractors using the transport map. For a color version of this figure, see www.iste.co.uk/gentil/geometric.zip*

COMMENTS ON FIGURE 1.19.– *On the left-hand side, the attractor of the first IFS is the segment $[0,1]$; on the right-hand side, the second attractor is defined in $\mathbb{R}^2$. For the segment, $\phi(10^\omega) = \phi(01^\omega)$ (green square on the left-hand side of the image), while for the left-hand attractor, $\phi'(10^\omega) \neq \phi'(01^\omega)$ (the two green squares on the right-hand side of the image). This mapping does not define a morphism from the first attractor onto the second.*

1.1.10. *Example of a curve*

To build a figure with a given topology, we start by defining an IFS whose attractor represents the parameter space. This IFS is a reference that we use to identify the multiple address points. In the second attractor, these addresses will have to reference the same point of the attractor.

Let us take an example of an IFS consisting of two transformations $\{T_0, T_1\}$. For a curve, we choose $[0, 1]$ as parameter space. By choosing $T_0(x) = \frac{1}{2}x$ and $T_1(x) = \frac{1}{2}x + \frac{1}{2}$, we have $\mathcal{A} = [0, 1]$ since $T_0([0, 1]) \cup T_1([0, 1]) = [0, 1]$. For this attractor, only one point has multiple addresses: $\frac{1}{2} = \phi(01^\omega) = \phi(10^\omega)$. To build a morphism that defines a fractal curve parameterized by $[0, 1]$, one just has to choose an IFS $\{T_0', T_1'\}$, such that:

$$T_1'(H(T_1^{-1}(1/2))) = T_0'(H(T_0^{-1}(1/2))) \qquad [1.2]$$

In fact, we have $T_1^{-1}(1/2) = 0 = \phi(0^\omega)$ (a fixed point of T_0) and $T_0^{-1}(1/2) = 1 = \phi(1^\omega)$ (a fixed point of T_1). We induce that $H(T_1^{-1}(1/2)) = H(\phi(1^\omega)) = \phi'(1^\omega)$ is the fixed point of T_1' and that $H(T_0^{-1}(1/2)) = H(\phi(0^\omega)) = \phi'(0^\omega)$ is the fixed point of T_0'. This brings us back to the condition:

$$T_1'(\phi'(0^\omega)) = T_0'(\phi'(1^\omega))$$

which we can write in the form:

$$T_1'(c_0') = T_0'(c_1')$$

with c_0' and c_1' the respective fixed points of T_0' and T_1'. This constraint corresponds to the connection constraint used for the construction of the FIF.

If the condition expressed by equation [1.2] is verified, the parameterized curve is defined by:

$$\forall t \in [0, 1] \; C(t) = \phi'(\phi^{-1}(t))$$

The computation of $\phi^{-1}(t)$ can be achieved by using the escape algorithm (see Barnsley (1988)).

1.1.11. *The example of the Sierpinski triangle*

Independent of the fractal topological structures being considered, we need to define a reference parameter space, built from an IFS, defining the topology of the fractal. For the Sierpinski triangle, we can choose as the IFS: $\{\mathbb{C}; \frac{z}{2}, \frac{z}{2} + \frac{1}{2}, \frac{z}{2} + i\}$. The associated attractor is the Sierpinski triangle defined in $\mathbb{C}$ and represented in Figure 1.20.

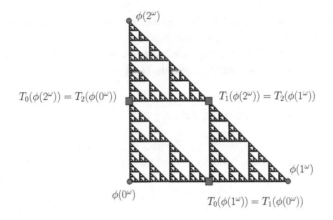

Figure 1.20. *Attractor defining the parameter space for the Sierpinski triangle built from the IFS $\{\mathbb{C}; \frac{z}{2}, \frac{z}{2} + \frac{1}{2}, \frac{z}{2} + i\}$. Double address points are indicated by green squares. They define the constraints that an IFS must meet in order to define a morphism from the Sierpinski triangle. For a color version of this figure, see www.iste.co.uk/gentil/geometric.zip*

We can then identify three double address points:

– the point $\frac{1}{2}$ of addresses 01^ω and 10^ω;

– the point $\frac{i}{2}$ of adresses 02^ω and 20^ω;

– the point $\frac{1}{2} + \frac{i}{2}$ of addresses 12^ω and 21^ω.

To build a morphism from the Sierpinski triangle, the second IFS must verify the constraints:

$$T'_0(\phi'(1^\omega)) = T'_1(\phi'(0^\omega))$$
$$T'_0(\phi'(2^\omega)) = T'_2(\phi'(0^\omega))$$
$$T'_1(\phi'(2^\omega)) = T'_2(\phi'(1^\omega))$$

REMARK.– This construction does not quite guarantee the equivalence of topologies. The condition defined on multiple addresses is a necessary but in sufficient condition for the transport mapping to be a homeomorphism, and guarantees the preservation of the topology. A reciprocal property should be added thereto: if a point of the second attractor has multiple addresses, these addresses must correspond to a single point for the first attractor.

REMARK.– The implementation of this construction is simple when the number of multiple address points is finite. This is the case for curves or wired structures. For surfaces, structures such as the Menger sponge or volume structures, the number of multiple address points is infinite. In section 1.3.5, we shall present how to use the decomposition property of an attractor, into sub-attractors, to identify these points and guarantee the condition expressed by equation [1.1].

1.2. *Controlled Iterated Function System*

In an IFS, all transformations are applied at each iteration. It is possible to enhance this model by adding rules to control the iterations. This is the principle of a C-IFS. C-IFS are more general systems that allow us to control certain parts of the IFS attractor. A C-IFS refers to an IFS with restrictions on the transformation sequences imposed by a control graph. The notion of C-IFS is present in literature under different formulations. It is known as *recurrent IFS* (R-IFS) (Barnsley 1988), and is also described by means of formal languages, called LR-IFS (*language-restrict iterated function system*) (Prusinkiewicz and Hammel 1994; Thollot and Tosan 1993), as well as L-systems (Prusinkiewicz and Hanan 1990; Terraz *et al.* 2009).

The attractor of a C-IFS can be evaluated by an automaton (Mauldin and Williams 1988) defined in the control graph. Each accepted word of the automaton corresponds to an authorized composition of transformations. Each state of the automaton corresponds to different parts of the modeled object. The states are associated with the construction spaces. Transitions between states indicate that one subset is contained inside of another. It is then possible to control the attractor more accurately.

C-IFS defines objects whose geometry can be complex. However, C-IFS attractors are more practical and controllable than IFS attractors for manufacturing. The common idea of these different approaches is to break "strict" self-similarity by controlling the iterative system with rules, in order to generate a more significant variety of forms. These can be expressed with either grammars or automata, by generating a subset of authorized infinite words. Therefore, of all possible transformation combinations, only some will be "authorized". We will not systematically explore all branches of the evaluation tree, some will be deliberately omitted and the corresponding parts of the attractor will not be considered. The version that we will use, which is presented here, is that of control based on automatons.

To better understand this idea, consider an IFS $\mathbb{T}$ consisting of three transformations of $\mathbb{T} = \{T_i\}_{i \in \Sigma}$, $\Sigma = \{0, 1, 2\}$. Within the context of the theory of IFS, all combinations are allowed. The systematic construction of all possible combinations can be represented by a single-state automaton (the initial state denoted by ♮) and a set of transitions (one by transformation) ending at that same state (see Figure 1.21). The words accepted by the automaton are by definition, the addresses of the points of the attractor, which correspond in our Σ^ω to the set of all the infinite words of Σ. The attractor associated with this automaton will then be $A(\mathbb{T})$.

REMARK.– We distinguish between the transitions of associated transformations, since they are two different mathematical entities. However, a semantic link will be induced by naming each transition according to the index of the associated transformation: transition σ is associated with transformation T_σ. For the sake of clarity of patterns, the representation of several transitions between the same two states will be done using a single arrow between states. This arrow will be labeled with the list of transition names separated by a comma, namely, the list of transformation indices. This notation will evolve slightly in the following to enhance semantics.

Figure 1.21. *Automaton of an IFS* $\mathbb{T} = \{T_0, T_1, T_2\}$. *The transition i is associated with the transformation of T_i. The state ♮ represents the initial state*

Now, let us consider the automaton represented by Figure 1.22. The words generated by this automaton are either of the form $\sigma = 0\sigma_{i_0} \cdots \sigma_{i_n} \cdots$ with $\sigma_{i_n} \in \{0, 2\}$ or of the form $\sigma' = 2\sigma_{j_0} \cdots \sigma_{j_m} \cdots$ with σ_{j_m} with $\sigma_{j_n} \in \{0, 1\}$. These two types of addresses correspond either to the addresses of the points of the sub-attractor of the IFS $\{T_0, T_2\}$ prefixed by 0, or to the addresses of the points of the sub-attractor of the IFS $\{T_0, T_1\}$ prefixed by 2. The attractor associated with the initial state is then composed of the union of $T_0(A(\{T_0, T_2\}))$ and $T_2(A(\{T_0, T_1\}))$. For this example, the state A (respectively, B) corresponds to the automaton of the IFS $\{T_0, T_2\}$ (respectively, $\{T_0, T_1\}$).

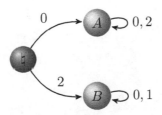

Figure 1.22. *Example of a three-state automaton inducing a restriction of the set of accepted words with respect to the automaton shown in Figure 1.21. For a color version of this figure, see www.iste.co.uk/gentil/geometric.zip*

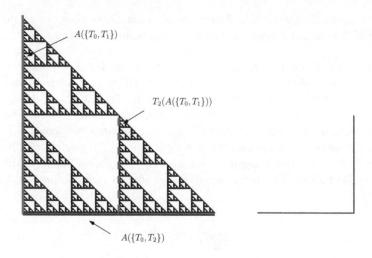

Figure 1.23. *Both images represent the attractors defined from the same transformation set of* $\mathbb{C}$*:* $T_0 = \frac{z}{2}$*,* $T_1 = \frac{z}{2} + \frac{i}{2}$ *and* $T_2 = \frac{z}{2} + \frac{1}{2}$*.*
For a color version of this figure, see www.iste.co.uk/gentil/geometric.zip

COMMENTS ON FIGURE 1.23.– *The image on the left corresponds to the attractor of the automaton in Figure 1.21, the one on the right side in Figure 1.22. On the left side, the red-colored segment that is a solid line represents the attractor* $\mathcal{A}(\{T_0, T_1\})$ *and the red dotted line* $T_2(\mathcal{A}(\{T_0, T_1\}))$*. The blue segment represents the attractor* $\mathcal{A}(\{T_0, T_2\})$*. On the right side, the attractor is the union of* $T_2(\mathcal{A}(\{T_0, T_1\}))$ *with* $T_0(\mathcal{A}(\{T_0, T_2\}))$*.*

Figure 1.24. *Other examples of attractors built from the same automatons as those in Figure 1.23 but with another set of transformations*

As we shall specify hereafter, each state is associated with an attractor. As shown in Figure 1.22, from the state A, the automaton generates the words of $\{0, 2\}^\omega$. The attractor associated with the state A is $\mathcal{A}(\{T_0, T_2\})$. Similarly, the attractor associated with state B is $\mathcal{A}(\{T_0, T_1\})$ (see Figures 1.23 and 1.24). We therefore see how we can define combination operations of attractors from combinations of automatons. If a state E belongs to the automaton of an IFS, we understand that the attractor of that IFS is a sub-part of the attractor of the automaton. In the overall attractor, this subpart appears transformed by the set of transformations corresponding to the path traveling from the initial state to the state E. In particular, if this state is connected to the initial state by a transition whose associated transformation is the identity, the attractor will appear unchanged in the final attractor. For example, the construction of the union of two attractors $\mathcal{A}(\mathbb{T})$ and $\mathcal{A}(\mathbb{T}')$ can be described using the automaton given by Figure 1.25. Let us assume that $\mathbb{T}' = \{T_i\}_{i\in\Sigma}$ and $\mathbb{T} = \{T'_{i'}\}_{i'\in\Sigma'}$. We define two states A and B, one for each IFS, each with their associated transitions. We add an initial state, with a transition to each state A and B. Each of these two transitions is associated with the identity transformation. The words generated by this automaton are in the form 0σ, where $\sigma \in \Sigma^\omega$, or of the form $0'\sigma'$, where $\sigma' \in \Sigma'^\omega$. Since we generate all the addresses of the points of $\mathcal{A}(\mathbb{T})$ preceded by the index 0, representing the identity, the final result contains $\mathcal{A}(\mathbb{T})$. Similarly, the other addresses correspond to the addresses of the points of $\mathcal{A}(\mathbb{T}')$ preceded by the index $0'$, itself also representing the identity, thus the final result also contains $\mathcal{A}(\mathbb{T}')$. The attractor associated with Figure 1.25 (with the identity associated with transitions 0 and $0'$) is the union of the attractors of states A and B. In order to be in accordance with our notations, we assume that $\Sigma \cap \Sigma' = \emptyset$, $0 \notin \Sigma$ and $0' \notin \Sigma'$.

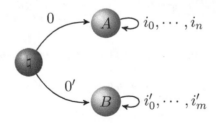

Figure 1.25. *Automaton generating the union of two attractors. Transitions 0 and $0'$ are associated with the identity transformation. For a color version of this figure, see www.iste.co.uk/gentil/geometric.zip*

We present a final example showing the value of the automaton-based approach. This example corresponds to the result that we can obtain by intersection of a Menger sponge with a particular plane (see Figure 1.26). The result is, overall, a hexagonal shape. However, it is not self-similar in the sense that we defined in section 1.1.2. To describe this shape, two subdivision processes have to be involved. The hexagonal shape presents a star in the center. We notice that this star is reproduced in six copies on a smaller scale around the center, represented schematically by hexagonal shapes of blue in Figure 1.27. This self-similarity is described by defining a state representing this hexagonal form with six transitions ending at this same state. These transitions reflect the subdivision/decomposition of the hexagonal form into six parts of the same nature, but smaller. Nonetheless, we have not fully described the form. Between these hexagonal parts appear triangular-shaped parts, as shown in green in Figure 1.27. We then define a second state for this triangular shape. We add six transitions going from the hexagonal state to the triangle state, specifying that the hexagonal shape is also subdivided into six triangular shapes. We still need to define the subdivision of the triangular shape to complete the description. Analogously to the hexagonal shape, the triangular shape is decomposed into three triangles and one hexagon. We then complete the automaton by adding four transitions: three ending at the triangle state and one going from the triangle state to the hexagon state. The initial state is the hexagon state.

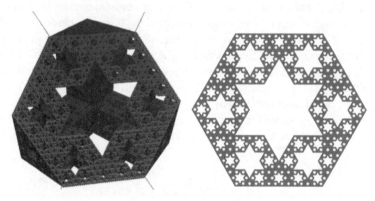

Figure 1.26. *The internal structure of the Menger sponge. On the left, the intersection of the Menger sponge with a half-space. On the right, a representation of this intersection with the plane defining the half-space. For a color version of this figure, see www.iste.co.uk/gentil/geometric.zip*

REMARK.– In the examples given above, a number of elements are implicit and need to be specified. Others may be more general than these examples suggest, in particular:

– attractors associated with each state are not necessarily defined in the same spaces, as we will see later;

– the direction of the transitions corresponds to the opposite direction of the application of transformations. For example, in Figure 1.27, transition 6 going from the state "hexagon" to the state "triangle" represents the action of the transformation T_6 applied to the attractor of the triangle state, that is, $T_6(\triangle)$. The attractor associated with the triangle state is transformed by the transformation T_6 to be copied into the attractor of the hexagonal state.

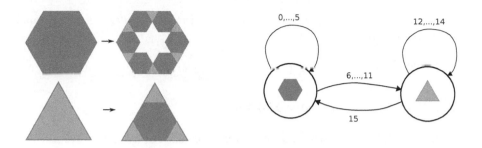

Figure 1.27. *Automaton describing the structure of the image on the right-hand side in Figure 1.26. For a color version of this figure, see www.iste.co.uk/gentil/geometric.zip*

1.2.1. *Formalization*

In this section, we give the formal definitions of C-IFS and associated attractors.

DEFINITION.– *A C-IFS is given by:*

– *an automaton* $(\Sigma, Q, \delta, \natural)$ *where* Σ *is an alphabet,* Q *a set of states,* δ *a transition function* $\delta : Q \times \Sigma \to Q$ *and* $\natural$ *the initial state;*

– *a set of states associated with spaces:* $(\mathbb{X}^q)_{q \in Q}$;

– *a set of operators associated with transitions* $T_i^x : E^{\delta(x,i)} \to E^x$;

– *a set of display primitives associated with states:* $(K^q)_{q \in Q}$ *where* $K^q \in \mathcal{H}(\mathbb{X}^q)$. *These primitives are utilized as initial primitives for each state in order to evaluate the attractor but are not used to define the attractor.*

We denote the restriction of Σ to the outgoing transitions associated with the state q by Σ^q, that is:

$$\Sigma^q = \{i \in \Sigma, \ \delta(q,i) \in Q\}$$

DEFINITION.– *Given a C-IFS, each state* $q \in Q$ *is associated with an attractor denoted by* $\mathcal{A}^q$:

$$\mathcal{A}^q = \bigcup_{i \in \Sigma^q} T_i^q(\mathcal{A}^{\delta(q,i)}) \qquad [1.3]$$

Each attractor $\mathcal{A}^q$ belongs to $\mathcal{H}(\mathbb{X}^q)$. They are mutually recursive, as we have seen in the example given in Figure 1.27.

DEFINITION.– *The attractor associated with a C-IFS is the attractor of the initial state, in other words:*

$$A = \mathcal{A}^\natural$$

1.2.2. *Visualization and approximation*

Similar to IFS, the attractor associated with each state of a C-IFS is evaluated by constructing a sequence of objects $\{K_n^q\}_{n\in\mathbb{N}}$ converging toward the attractor $\mathcal{A}^q$:

$$
\begin{aligned}
K_0^q &= K^q \\
K_{n+1}^q &= \bigcup_{i\in\Sigma^q} T_i^q(K_n^{\delta(q,i)})
\end{aligned}
$$

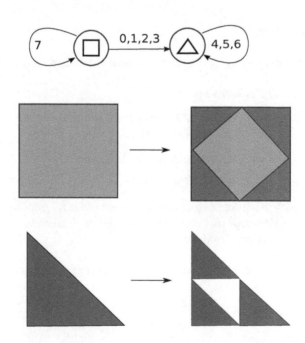

Figure 1.28. *Example of a two-state automaton: the □ is divided into four △ and one □, the △ is divided into three △. For a color version of this figure, see www.iste.co.uk/gentil/geometric.zip*

Let us take a new example whose automaton is represented by Figure 1.28. This attractor is built from two subdivision rules, each corresponding to an automaton state: a state □ (square) and a state △ (triangle), due to the display primitives associated with them. The square is decomposed into one square and four triangles. The triangle is decomposed into three triangles. The transition function is then:

$$\delta(\square, 0) = \triangle \quad \delta(\triangle, 4) = \triangle$$
$$\delta(\square, 1) = \triangle \quad \delta(\triangle, 5) = \triangle$$
$$\delta(\square, 2) = \triangle \quad \delta(\triangle, 6) = \triangle$$
$$\delta(\square, 3) = \triangle$$
$$\delta(\square, 7) = \square$$

The initial state $\natural = \square$ and the evaluation of $\mathcal{A}^{\natural}$ is done as follows:

– *Case of zero iterations (K_0)*: we start from the initial state, as there is no iteration (no subdivision), the approximation is made from the primitive associated with the current state, that is, the initial state. We get K_0:

$$K_0 = K^{\square}$$

– *Case of one iteration (K_1)*: from the original state, we apply each authorized subdivision. Therefore, for each transition leaving the initial state, we apply the transformation (associated with the transition) to the primitive of the arrival state. We obtain $K_1^{\square}$:

$$K_1 = \bigcup_{i \in \Sigma} T_i(K_0^{\delta(\square, i)})$$
$$= T_0(K^{\triangle}) \cup T_1(K^{\triangle}) \cup T_2(K^{\triangle}) \cup T_3(K^{\triangle}) \cup T_7(K^{\square})$$

– *Case of two iterations (K_2)*: we apply the same algorithm for each arrival state from the previous state

$$K_2 = T_0 T_4(K^{\triangle}) \cup T_0 T_5(K^{\triangle}) \cup T_0 T_6(K^{\triangle})$$
$$\cup T_1 T_4(K^{\triangle}) \cup T_1 T_5(K^{\triangle}) \cup T_1 T_6(K^{\triangle})$$
$$\cup T_2 T_4(K^{\triangle}) \cup T_2 T_5(K^{\triangle}) \cup T_2 T_6(K^{\triangle})$$
$$\cup T_3 T_4(K^{\triangle}) \cup T_3 T_5(K^{\triangle}) \cup T_3 T_6(K^{\triangle})$$
$$\cup T_7 T_0(K^{\triangle}) \cup T_7 T_1(K^{\triangle}) \cup T_7 T_2(K^{\triangle}) \cup T_7 T_3(K^{\triangle}) \cup T_7 T_7(K^{\square})$$

Figure 1.29 shows the first four approximations of the automaton attractor shown in Figure 1.28. The evaluation tree developed at level 2 is presented in Figure 1.31. The approximation at level 9 is presented in Figure 1.30.

1.2.3. *Implementation*

As with IFS, it is possible to utilize the two types of algorithms of section 1.1.6. We only present the adaptation of the first (it is similar for the second). This modification simply consists of not systematically applying all transformations but only those associated with the transitions of the current state.

Figure 1.29. *Construction of the sequence converging to the attractor of the automaton of Figure 1.28 (from the top to the bottom and from the right to the left: $K_0^{\square}$, $K_1^{\square}$, $K_2^{\square}$ and $K_3^{\square}$). For a color version of this figure, see www.iste.co.uk/gentil/geometric.zip*

Figure 1.30. *Approximation of the automaton attractor of Figure 1.28 obtained with nine iterations*

1.3. *Boundary controlled iterated function system*

With the IFS and C-IFS models, we can very easily generate complex structures. The attractors constantly depend on transformations (Barnsley 1988). However, this continuity is established for the topology induced by the Hausdorff distance; it does not guarantee topological equivalence between two attractors, as close as they are.

From the designer's point of view, this is very penalizing. Indeed, even a tiny variation of transformation parameters can generate an attractor whose topological structure is radically different. A volume can be transformed into a curve or even a set of accumulation points (homeomorphic to a Cantor set).

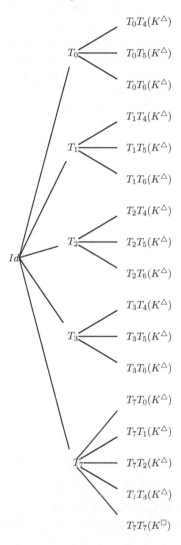

$T_0T_4(K^\triangle)$

T_0 — $T_0T_5(K^\triangle)$

$T_0T_6(K^\triangle)$

$T_1T_4(K^\triangle)$

T_1 — $T_1T_5(K^\triangle)$

$T_1T_6(K^\triangle)$

$T_2T_4(K^\triangle)$

T_2 — $T_2T_5(K^\triangle)$

Id

$T_2T_6(K^\triangle)$

$T_3T_4(K^\triangle)$

T_3 — $T_3T_5(K^\triangle)$

$T_3T_6(K^\triangle)$

$T_7T_0(K^\triangle)$

$T_7T_1(K^\triangle)$

T_7 — $T_7T_2(K^\triangle)$

$T_iT_3(K^\triangle)$

$T_7T_7(K^\square)$

Figure 1.31. *Evaluation tree developed at level 2, for the attractor of the automaton given by Figure 1.28*

On the other hand, describing a form using mathematical transformations is a difficult academic exercise and not always intuitive, especially for users unfamiliar with mathematics. For others, this often requires a learning and experimentation phase. It is, however, also possible to develop user interfaces that facilitate the specification of transformations by hiding the "mathematical parameters". We can also rely on theoretical results, such as the collage theorem by Barnsley (1988), to facilitate the description of self-similarity of a given form. In any case, however, the problem of the non-preservation of topology when changing parameters remains an obstacle to the control of forms and, thus, to creativity. In addition, graphic designers are familiar with a set of tools that assist them in editing and manipulating shapes. For example, most surfaces (NURBS or subdivision surfaces) are described from a grid of control points. The user distorts and adjusts the geometry of the surface from these control points without worrying about mathematical functions that are not directly accessible (see Figure 1.32).

Ensure: $K_n = \bigcup_{\sigma \in \Sigma} T_\sigma(K_{\sigma_n})$
Require: $\{T_i\}_{i \in \Sigma},\, n > 0,\, (\Sigma, Q, \delta, \natural)$

$\quad T \leftarrow Id_{\mathbb{X}^\natural}$ ▷ Identity transformation of $\mathbb{X}^\natural$

$\quad$ Display($T,n,\natural$)

$\quad$ **procedure** DISPLAY(T,m,q)

$\quad\quad$ **if** $m = 0$ **then**

$\quad\quad\quad K_n \leftarrow T(K^q)$

$\quad\quad\quad$ Display K_n

$\quad\quad$ **else**

$\quad\quad\quad$ **for** $i \in \Sigma^q$ **do**

$\quad\quad\quad\quad T \leftarrow T \circ T_i^q$

$\quad\quad\quad\quad$ Display($T,m-1,\delta(q,i)$)

$\quad\quad\quad$ **end for**

$\quad\quad$ **end if**

$\quad$ **end procedure**

Algorithm 1.2. *Visualization K^n*

The model that we present in this section aims to define and control the topology of self-similar objects and to modify the geometry using control points while preserving the topology.

To represent geometric shapes, one conventionally separates the topological description of the shape from its geometric description (what we call geometric embedding). There are multiple advantages, as follows: the data structure is independent of the geometry, the notion of neighborhood is explained, manipulation is more efficient, etc. This approach is even more relevant for fractal forms because, in addition to the aforementioned advantages, it allows us to highlight two types of "fractalities": topological fractality and geometric fractality. By dissociating them from the representation model, they can be independently manipulated and controlled. A surface tile B-spline does not have topological fractality (it is a surface of topological dimension 2), nor geometric fractality (it is a smooth surface). The Menger carpet, traditionally represented in the plane, can also be built within a 3D space. It has a topology of the fractal type, its topological dimension is equal to 1, but is neither homeomorphic to a curve nor to a surface. Nonetheless, its geometry can be smooth (see Figure 1.33, top right). The surface tile B-spline can be modified into a rough surface, and its geometry is then fractal because its fractal dimension is greater than its topological dimension, but there still remains a surface (see Figure 1.33 bottom left). Finally, the Menger tile can undergo the same type of modification, to give it a rough appearance. It then acquires the two natures of fractality: topological and geometric (see Figure 1.33, bottom right).

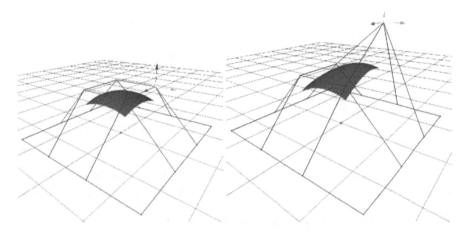

Figure 1.32. *Example of a third-degree B-spline surface defined from a grid of control points. Modifying the position of the control points impacts the geometry of the surface. On the left, a bicubic B-spline surface defined from a grid of control points. On the right, a control point was moved, thus changing the geometry of the surface. For a color version of this figure, see www.iste.co.uk/gentil/geometric.zip*

Describing self-similar shapes is more complex than describing surfaces. A surface corresponds to a specific, well-defined topological structure. Self-similar forms may have very varied topological structures (curves, surfaces, volumes, wired structures, loose density structures, etc.). In addition, fractal topological structures are not clearly identified and listed and there are an infinite number of them. However, we shall see that it is possible to define self-similar topological structures as equivalence classes, defined by a set of incidence and adjacency relations between the parts that make up the form.

Figure 1.33. *The surface, at the top right, is a smooth B-spline surface and has neither topological fractality nor geometric fractality. For a color version of this figure, see www.iste.co.uk/gentil/geometric.zip*

COMMENTS ON FIGURE 1.33.– *In the bottom left, the surface is rough and presents only geometric fractality. In the top right, the Menger carpet has undergone a polynomial deformation and presents only topological fractality. Finally, in the bottom right image, the Menger carpet presents roughness. It has both topological and geometric fractality.*

This principle has been used in the context of fractal interpolation functions (FIF) for curves and surfaces (Barnsley 1986). For curves, the principle consists of connecting the parts of the attractor to one another by a point, ensuring continuity between these parts, as illustrated in Figure 1.34.

We can see the BC-IFS model as a generalization of FIF with three evolutions.

1) The fittings are defined, not only by points (for curves) or curves (for surfaces), but by sub-attractors of any topological nature.

2) In order to be able to deform the attractors intuitively, as for NURBS surfaces or subdivision surfaces, attractors are built in barycentric spaces, similarly to the basis functions of NURBS. Each attractor point represents a weighted sum of a set of control points. The attractor is then projected onto the modeling space, depending on the positions of the control points. The shift of control points in the modeling space generally changes the shape of the object, but its nature remains the same, regardless of its initial topological and geometric characteristic s.

3) We introduce the possibility of controlling the local aspect of shapes, that is, geometric fractality. This is done by introducing the concept of "subdivision points". In the case of subdivision surfaces and, for some NURBS, surface visualization

algorithms, subdivision points are implicit and fixed by the basis functions of the models. We shall explain this notion to obtain finer control and generate more varied geometries. These subdivision points allow us to modify the basis functions and control the geometric appearance (smooth or rough).

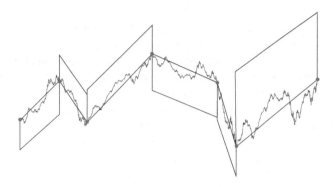

Figure 1.34. *Example of a curve constructed based on an FIF. The parallelepipeds symbolize the self-similar parts, that is, the images of the curve by each of the transformations. The dots in red are the interpolated points according to which the parts are connected. For a color version of this figure, see www.iste.co.uk/gentil/geometric.zip*

1.3.1. *Attractors defined from a set of control points*

Curves and NURBS surfaces are defined from a set of control points and basis functions. The parametric formulation for curves is given as:

$$C(t) = \sum_{i=0}^{m-1} P_i B_i(t), \ t \in [0, 1] \tag{1.4}$$

where t represents the parameter, $[0, 1]$ is the parameterization domain, $C(t)$ the point of the curve for the value of the parameter t, P_i the ith control point and $B_i(t)$ the ith basis function. The basis functions must verify the property: $\Sigma_{i=0}^{m-1} B_i(t) = 1$, $\forall t \in [0, 1]$. This property ensures that equation [1.4] is well-defined and represents a weighted sum of control points, which is a barycentric combination. Weight values are determined for each value of parameter t using the basis functions.

In the case of quadratic Bezier curves, the basis functions are given as:

$$B_0(t) = (1 - t)^2$$
$$B_1(t) = 2t(1 - t)$$
$$B_2(t) = t^2$$

Thus, $B(t) = \begin{pmatrix} B_0(t) \\ B_1(t) \\ B_2(t) \end{pmatrix}, t \in [0, 1]$ defines a curve in the barycentric space of

dimension 3, denoted by $BI^3 = \{\lambda = \begin{pmatrix} \lambda_0 \\ \lambda_1 \\ \lambda_2 \end{pmatrix} \in \mathbb{R}^3 | \lambda_0 + \lambda_1 + \lambda_2 = 1\}$. The projection

onto the modeling space is achieved by simply achieving the barycentric combination of control points for each value of the parameter t. If we move the control points, the curve is modified according to the new positions, but the base functions remain the same. The nature of the curve is defined by the basis functions and the overall shape is controlled by the positions of the control points. Users choose the type of curve (which implies specific base functions) without worrying about the equations, and adjust the shape of their curve using control points. Another advantage is that the size of the modeling space ($\mathbb{R}^2$, $\mathbb{R}^3$ or $\mathbb{R}^n$) only affects the number of coordinates of the control points. On the other hand, the size of the barycentric space depends on the number of control points desired or necessary to distort the curve.

We thus make a distinction between three spaces: the parameter space or parameterization domain, the modeling space in which we want to define the curve and the barycentric space in which the basis functions are defined.

Similarly, we build the attractors in a barycentric space. This attractor is then projected onto the modeling space according to the control points.

1.3.2. *Formulation*

DEFINITION.– *A barycentric space of dimension m is a sub-space of $\mathbb{R}^m$, denoted BI^m and defined by:*

$$BI^m = \{\lambda = \begin{pmatrix} \lambda_0 \\ \lambda_1 \\ \vdots \\ \lambda_{m-1} \end{pmatrix} \in \mathbb{R}^m | \sum_{i=0}^{m-1} \lambda_i = 1\}$$

BI^2 is the line of $\mathbb{R}^2$ passing through the points $\begin{pmatrix} 1 \\ 0 \end{pmatrix}$ and $\begin{pmatrix} 0 \\ 1 \end{pmatrix}$. BI^3 is the plane

of $\mathbb{R}^3$ passing through points $\begin{pmatrix} 1 \\ 0 \\ 0 \end{pmatrix}$, $\begin{pmatrix} 0 \\ 1 \\ 0 \end{pmatrix}$ and $\begin{pmatrix} 0 \\ 0 \\ 1 \end{pmatrix}$. It is worth noting that the

components λ_i of a barycentric space point can be negative.

To define an IFS attractor, we saw, in section 1.1.4, that the first condition is to operate in a complete metric space. Barycentric spaces, seen as sub-spaces of $\mathbb{R}^m$, equipped with the Euclidian distance, are complete metric spaces. In addition, the transformations used to describe self-similarity must be internal, that is to say, the image of a point of the barycentric space must be a point of that same space. For the sake of simplicity, as with IFS and C IFS, we shall use linear transformations. Among all the linear transformations of $\mathbb{R}^m$ in $\mathbb{R}^m$, we are interested in those whose matrix representation is of the "stochastic matrix" type, but whose coefficients might be negative.

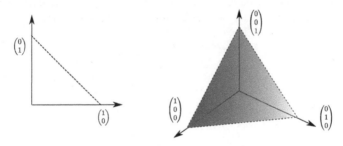

Figure 1.35. *Barycentric space. On the left, the barycentric space of dimension 2 is the line passing through the points $\begin{pmatrix} 1 \\ 0 \end{pmatrix}$ and $\begin{pmatrix} 0 \\ 1 \end{pmatrix}$. On the right, the barycentric space of dimension 3 is the plane passing through the points $\begin{pmatrix} 1 \\ 0 \\ 0 \end{pmatrix}$, $\begin{pmatrix} 0 \\ 1 \\ 0 \end{pmatrix}$ and $\begin{pmatrix} 0 \\ 0 \\ 1 \end{pmatrix}$. For a color version of this figure, see www.iste.co.uk/gentil/geometric.zip*

DEFINITION.– *We define as "barycentric transformation of BI^m" any linear transformation of $\mathbb{R}^m$ in $\mathbb{R}^m$, whose matrix representation $(t_{i,j})_{i,j \in \{0,\cdots,m-1\}}$ in the canonical base verifies $\forall j \in \{0, \cdots, m-1\}$, $\Sigma_i t_{i,j} = 1$.*

PROPERTY.– Let T be a barycentric transformation of BI^m and $(t_{i,j})_{i,j \in \{0,\cdots,m-1\}}$ its matrix representation in the canonical base, then $\forall \lambda \in BI^m$, $T(\lambda) \in BI^m$.

Finally, the last condition (sufficient but not necessary) for the existence of a single attractor associated with an IFS is that each transformation of the IFS be contractive. This condition is reflected by the fact that the eigenvalues of the matrix must be strictly smaller than 1.

PROPERTY.– Let T be a barycentric transformation of BI^m and M_T its matrix representation in the canonical base, then M_T has 1 as its eigenvalue.

PROOF.– Same demonstration as for stochastic matrices.

Let T be a barycentric transformation of BI^m and M_T its matrix representation in the canonical base, then T is contractive if all the eigenvalues of M_T, other than the eigenvalue 1, are of modulus strictly less than 1.

EXAMPLES.– The definition of an attractor in a barycentric space is always done by translating self-similarities, but this time with the help of barycentric transformations. The identification of the transformations is simply done. To obtain the matrix representation of a transformation, one simply has to express the coordinates of the transformations of the points of the canonical base in this same base:

$$
\begin{array}{c}
\begin{array}{ccccc} T(e_1) & \cdots & T(e_i) & \cdots & T(e_m) \end{array} \\
\begin{array}{c} e_1 \\ \vdots \\ e_i \\ \vdots \\ e_m \end{array}
\left(
\begin{array}{ccccc}
\bullet & \cdots & \bullet & \cdots & \bullet \\
\vdots & \cdots & \vdots & \cdots & \vdots \\
\bullet & \cdots & \bullet & \cdots & \bullet \\
\vdots & \cdots & \vdots & \cdots & \vdots \\
\bullet & \cdots & \bullet & \cdots & \bullet
\end{array}
\right)
\end{array}
$$

where e_i represents the barycentric point of coordinates $(\delta_{i,0} \cdots \delta_{i,j} \cdots \delta_{i,m})^T$, with $\delta_{i,j} = 1$ if $i = j$, and 0 otherwise.

1.3.2.1. *The example of the Cantor set*

The IFS $\{T_0, T_1\}$ with $T_0 = \begin{pmatrix} 1 & \frac{2}{3} \\ 0 & \frac{1}{3} \end{pmatrix}$ and $T_1 = \begin{pmatrix} \frac{1}{3} & 0 \\ \frac{2}{3} & 1 \end{pmatrix}$ has as attractor, the Cantor set, built between the points $\begin{pmatrix} 1 \\ 0 \end{pmatrix}$ and $\begin{pmatrix} 0 \\ 1 \end{pmatrix}$, as illustrated in Figure 1.36.

1.3.2.2. *The example of the Sierpinski triangle*

The IFS $\{T_0, T_1, T_2\}$ with $T_0 = \begin{pmatrix} 1 & \frac{1}{2} & \frac{1}{2} \\ 0 & \frac{1}{2} & 0 \\ 0 & 0 & \frac{1}{2} \end{pmatrix}$, $T_1 = \begin{pmatrix} \frac{1}{2} & 0 & 0 \\ \frac{1}{2} & 1 & \frac{1}{2} \\ 0 & 0 & \frac{1}{2} \end{pmatrix}$ and

$T_2 = \begin{pmatrix} \frac{1}{2} & 0 & 0 \\ 0 & \frac{1}{2} & 0 \\ \frac{1}{2} & \frac{1}{2} & 1 \end{pmatrix}$ has as attractor a Sierpinski triangle built between the points $\begin{pmatrix} 1 \\ 0 \\ 0 \end{pmatrix}$, $\begin{pmatrix} 0 \\ 1 \\ 0 \end{pmatrix}$ and $\begin{pmatrix} 0 \\ 0 \\ 1 \end{pmatrix}$, as illustrated in Figure 1.37.

Any point of the attractor can be considered as a set of coefficients of a weighted sum of m control points. The projection of the attractor into the modeling space simply consists of associating, at each point of the attractor, the point obtained by weighted sum of the control points (exactly as we do for NURBS curves and surfaces, from the basis functions, see Figure 1.38).

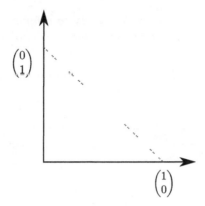

Figure 1.36. *Cantor set built in the barycentric space BI^2 using the IFS composed of transformations $T_0 = \begin{pmatrix} 1 & \frac{2}{3} \\ 0 & \frac{1}{3} \end{pmatrix}$ and $T_1 = \begin{pmatrix} \frac{1}{3} & 0 \\ \frac{2}{3} & 1 \end{pmatrix}$*

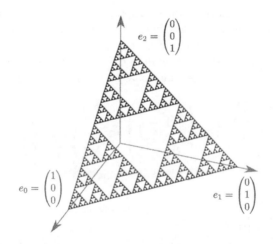

Figure 1.37. *Sierpinski triangle built in the barycentric space BI^3. For a color version of this figure, see www.iste.co.uk/gentil/geometric.zip*

DEFINITION.– *Let A be the attractor of an IFS defined in a barycentric space BI^m. The projection of A, according to the set of control points $\{P_0, \cdots, P_{m-1}\}$, is denoted by PA and defined by:*

$$PA = \{p \in \mathbb{X} \mid p = \Sigma_{i=0}^{m-1} \lambda_i P_i, \text{ with } \lambda = \begin{pmatrix} \lambda_0 \\ \lambda_1 \\ \vdots \\ \lambda_{m-1} \end{pmatrix} \in A\}$$

where $P = (P_0 \cdots P_i \cdots P_{m-1})$ *is the matrix of control points* (P_i *represents a column vector composed of the coordinates of the ith control point*).

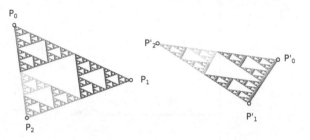

Figure 1.38. *Example of projections of the Sierpinski triangle. The attractor is defined based on a single IFS (see Figure 1.37), and two projections are calculated from two different sets of control points:* $\{P_0, P_1, P_2\}$ *and* $\{P'_0, P'_1, P'_2\}$. *For a color version of this figure, see www.iste.co.uk/gentil/geometric.zip*

This construction, presented for the IFS, applies equally well to C-IFS. Figure 1.39 presents a two-state automaton describing two subdivision systems. The first state $\square$ is divided into three states $\square$ and a state $\triangle$. The state $\triangle$ is divided into three states $\triangle$ as a Sierpinski triangle. The state $\square$ is built in BI^4 and the state $\triangle$ in BI^3. The matrix expressions of barycentric transformations are given as:

$$T_0 = \begin{pmatrix} 1 & \frac{1}{2} & \frac{1}{4} & \frac{1}{2} \\ 0 & \frac{1}{2} & \frac{1}{4} & 0 \\ 0 & 0 & \frac{1}{4} & 0 \\ 0 & 0 & \frac{1}{4} & \frac{1}{2} \end{pmatrix} \quad T_1 = \begin{pmatrix} \frac{1}{2} & 0 & 0 & \frac{1}{4} \\ \frac{1}{2} & 1 & \frac{1}{2} & \frac{1}{4} \\ 0 & 0 & \frac{1}{2} & \frac{1}{4} \\ 0 & 0 & 0 & \frac{1}{4} \end{pmatrix} \quad T_2 = \begin{pmatrix} \frac{1}{4} & 0 & 0 & 0 \\ \frac{1}{4} & \frac{1}{2} & 0 & 0 \\ \frac{1}{4} & \frac{1}{2} & 1 & \frac{1}{2} \\ \frac{1}{4} & 0 & 0 & \frac{1}{2} \end{pmatrix}$$

$$T_3 = \begin{pmatrix} 0 & \frac{1}{2} & 0 \\ 0 & 0 & 0 \\ 0 & 0 & \frac{1}{2} \\ 1 & \frac{1}{2} & \frac{1}{2} \end{pmatrix}$$

The transformations T_4, T_5 and T_6 correspond to the transformations defining the Sierpinski triangle (see the example of the Sierpinski triangle above).

The projection of the attractor onto the modeling space can also be managed by the automaton. One simply has to add a state with associated modeling space and then a transition going from this new state to the state representing the attractor in the barycentric space. We associate the matrix whose columns correspond to the coordinates of the control points to this transition. This matrix can be seen as a projection matrix from the barycentric space to the modeling space. The complete automaton representing the projection in Figure 1.40 is given by Figure 1.41.

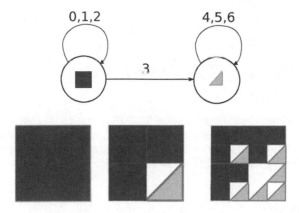

Figure 1.39. *Example of a two-state automaton. The □ is divided into three □ and a △. The △ is divided into three △. For a color version of this figure, see www.iste.co.uk/gentil/geometric.zip*

Figure 1.40. *Three different projections of the attractor described by the automaton in Figure 1.39 according to the set of control points $\{P_0, P_1, P_2, P_3\}$. On the left, $P_0 = (0,0)^\mathsf{T}$, $P_1 = (0,1)^\mathsf{T}$, $P_2 = (1,1)^\mathsf{T}$, $P_3 = (1,0)^\mathsf{T}$; in the middle, $P_0 = (0,0.2)^\mathsf{T}$, $P_1 = (0,0.8)^\mathsf{T}$, $P_2 = (0.7,1)^\mathsf{T}$, $P_3 = (1,0)^\mathsf{T}$; on the right, $P_0 = (0,0)^\mathsf{T}$, $P_1 = (0.1,1)^\mathsf{T}$, $P_2 = (0.9,0.3)^\mathsf{T}$, $P_3 = (0.6,0)^\mathsf{T}$*

REMARK.– The spaces of the different states are not necessarily of the same nature and *a fortiori* not necessarily of the same dimension. In our last example, the initial state is associated with the modeling space (here $\mathbb{R}^2$), at state □ the space BI^4 and, finally, at state △ the space BI^3. As a result, the matrix representations of the transformations associated with the transitions between two of these different states (of different sizes) are not square. This is the case with the transformation $T_3 : BI^3 \mapsto BI^4$. This transformation embeds the attractor associated with the △ state into the space BI^4 to form part of the attractor associated with the □. The same is true for the projection matrix composed of control points.

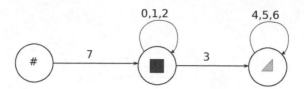

Figure 1.41. *Automaton defining the attractor in the barycentric spaces and performing the projection into the modeling space. The transformation associated with transition 7 and represented by the matrix composed of control points [P$_0$ P$_1$ P$_2$ P$_3$]. For a color version of this figure, see www.iste.co.uk/gentil/geometric.zip*

1.3.3. *Evaluation*

The evaluation of attractors, defined within barycentric spaces, is identical to that of attractors defined directly in the modeling space. It is done by iterations of the Hutchinson operator from the primitives $K^q, q \in Q$. Each primitive K^q must be defined in the space associated with the state q, that is, in a barycentric space. For example, for the evaluation of the Sierpinsky triangle shown in Figure 1.37, we chose

the triangle of vertices: $\begin{pmatrix} 1 \\ 0 \\ 0 \end{pmatrix}$ $\begin{pmatrix} 0 \\ 1 \\ 0 \end{pmatrix}$ and $\begin{pmatrix} 0 \\ 0 \\ 1 \end{pmatrix}$.

REMARK.– The philosophy of this formalism is to build an attractor in a barycentric space and then to project it into the modeling space, following a set of control points. An approximation of the attractor is calculated by iterations from the primitives associated with each state. This approximation is then projected by computing the barycentric combinations (the transformations of the primitives) of the control points. These calculations are only expressed by matrix products of the form $PT_{\sigma_0}T_{\sigma_1} \cdots T_{\sigma_i}K$. This view of the calculations is related to an interpretation of the matrix product from right to left. However, we can also interpret it from left to right. The product PT_{σ_0} is equivalent to applying n barycentric combinations (columns of T_{σ_0}) on the n control points stored in the P matrix. The matrix T_{σ_0} can then be seen as a transformation of the control points by left multiplication. Therefore, the application of a matrix T_j amounts to performing a refinement or subdivision of the structure of the control points. We shall see that we find the formalism of subdivision surfaces and Chaikin and De Casteljau algorithms for B-splines and Bezier surfaces exactly. This view is also useful for determining the coefficient values of subdivision matrices. When the number of control points is large, it is difficult to imagine barycentric spaces associated with attractors and to determine transformations that reflect self-similarity. One simply needs to reason on the transformations of the control points and express these images as the weighed sum of the initial control points. For example, by referring to Figure 1.39, the illustration of the first iterations of the attractor is not represented in BI^4, but is a projection following the control points positioned at the four vertices of the square. To determine the matrix T_3 by placing the $\triangle$ in the attractor $\square$, we notice that the first vertex of the triangle has the fourth control point as its image (first column of T_3: $(0, 0, 0, 1)^{\mathsf{T}}$), the second vertex has the middle of the first and fourth control points as its image (second column of T_3: $(\frac{1}{2}, 0, 0, \frac{1}{2})^{\mathsf{T}}$) and finally, the last control point has the middle of the third and fourth control points as its image (third column of T_3: $(0, 0, \frac{1}{2}, \frac{1}{2})^{\mathsf{T}}$).

1.3.4. Implementation

The implementation is rigorously identical to that of the C-IFS. Only the spaces change in nature, but the algorithms remain the same. There is therefore no adaptation of the algorithms to be made, apart from minimal adaptations of data structures, to define and manipulate transformations and points of spaces of any dimension ($\mathbb{R}^m$).

1.3.5. Topology control

Topological control is done by defining constraints on the transformations of the C-IFS. The transformations describe how the copies of the attractor are "placed" within the space. We shall require some copies of the attractor to have common subsets (as for the FIF, where two copies of a curve will share a common point to be connected). The sharing of subparts of the copies of the attractor is based on the lattice structure associated with the IFS that is transported to the attractors (see section 1.1.4).

Let two IFS, $\mathbb{T}_0 = \{T_0, T_1, T_2\}$ and $\mathbb{T}_1 = \{T_1, T_2, T_3\}$. These two IFS have a sub-IFS in common: $\{T_1, T_2\}$. This is reflected at the attractor level by a common sub-attractor: $\mathcal{A}(\{T_1, T_2\})$. Thus, independently of the transformations T_0, T_1, T_2 and T_3, $\mathcal{A}(\mathbb{T}_0)$ and $\mathcal{A}(\mathbb{T}_1)$ connect according to the sub-attractor $\mathcal{A}(\{T_1, T_2\})$ (see Figure 1.15). The attractors $\mathcal{A}(\mathbb{T}_0)$ and $\mathcal{A}(\mathbb{T}_1)$ may also have other intersection subsets, but they have at least this one. A particular case is where we choose a single transformation common to two IFS. The attractors will have (at least) one common point, the one corresponding to the fixed point of this transformation. We shall achieve these connections not between attractors but between parts of the same attractor to define a topological structure. A part of an attractor is the attractor itself, to which we have applied one of the transformations T_i. To connect the parts $T_i(\mathcal{A})$ and $T_j(\mathcal{A})$ by a common cell, we shall impose that $T_i C_x = T_j C_y$ where C_x and C_y are two sub-attractors (sub-IFS attractors) of $\mathcal{A}$.

To formalize this principle of connection, we introduce a more general concept of topological cell than that conventionally defined in B-rep models. In the same way that a face can be bordered by edges, a cell is bordered by a subcell. However, these cells can be geometric structures, and are not necessarily homeomorphic to a simplex. The "faces" can be Sierpinski triangles or Menger carpets, etc., and the "edges" can be

Cantor sets, etc. Nonetheless, due to the lattice structure associated with the attractors, consistency between the cell hierarchy is maintained.

We present the BC-IFS model within the context of an attractor built in a barycentric space and then projected onto the modeling space following a set of control points.

We define a very general cell. In our model, cells are used to identify part of the shape that, on the one hand, will depend on a certain number of control points and, on the other hand, will be used to define connections between the subdivisions of the form.

DEFINITION.– *A cell is the attractor of an IFS defined in a given barycentric space.*

EXAMPLE.– If the IFS consists of a single contractive transformation in BI^n, $n > 1$ then the associated cell is the fixed point of the transformation of the IFS.

A trivial case is where $n = 1$. In this case, $BI^1 = \{1\}$. The only internal transformation possible is the constant transformation which associates 1 with 1. The cell is then reduced to the point $\{1\}$.

EXAMPLE.– In BI^2, the IFS $\{T_0, T_1\}$, with $T_0 = \begin{pmatrix} 1 & \frac{1}{2} \\ 0 & \frac{1}{2} \end{pmatrix}$ and $T_1 = \begin{pmatrix} \frac{1}{2} & 0 \\ \frac{1}{2} & 1 \end{pmatrix}$ defines

the line segment connecting the point $\begin{pmatrix} 1 \\ 0 \end{pmatrix}$ to the point $\begin{pmatrix} 0 \\ 1 \end{pmatrix}$ as a cell.

EXAMPLE.– In BI^3, the IFS $\{T_0, T_1\}$, with $T_0 = \begin{pmatrix} 1 & \frac{1}{2} & \frac{1}{4} \\ 0 & \frac{1}{2} & \frac{1}{2} \\ 0 & 0 & \frac{1}{4} \end{pmatrix}$ and $T_1 = \begin{pmatrix} \frac{1}{4} & 0 & 0 \\ \frac{1}{2} & \frac{1}{2} & 0 \\ \frac{1}{4} & \frac{1}{2} & 1 \end{pmatrix}$,

defines the curve described by the second-degree Bernstein polynomials as a cell (see section 3.1).

REMARK.– As pointed out by previous examples, the dimensions of the barycentric spaces are independent of the topological dimension of cells. They depend only on the number of control points that will be used to make the projection into the modeling space. A point or curve (or any other cell) can be defined in a barycentric space of any dimension $n > 0$.

As with B-reps, we define a hierarchy between cells: a subcell is a part of another cell. This relationship is achieved by means of two concepts: that of incidence operator or boundary operator and that of adjacency relations.

DEFINITION.– *We define the incidence operator of a cell C_1 defined in BI^n to a cell C_2 defined in BI^m ($n \leq m$) as any embedding associated with every canonical point e_i^n of BI^n, a canonical point e_i^m of BI^m.*

The incidence operator merely maps the dimensions of the barycentric space of one cell with the dimensions of the barycentric space of another cell, but does not define any constraints between the attractors representing the cells. These incidence operators define the hierarchy of the space of one cell into the space of another cell.

We define the concept of incidence in a very general way, and it should be seen as a concept of inclusion.

DEFINITION.– *Let C_1 be a cell defined in BI^n by an IFS $\mathbb{T}^1$ and C_2 a cell defined in BI^m by an IFS $\mathbb{T}^2$ ($n \leq m$) and $\amalg$ an incidence operator from C_1 to C_2.*

  C_1 is said to be "boundary of" or "incident to" C_2 if $\amalg(\mathcal{A}(\mathbb{T}^1)) \subset (\mathcal{A}(\mathbb{T}^2))$.

REMARK.– The introduction of the incidence operator in this definition allows barycentric spaces to be mapped, in particular, to formalize the fact that the attractor of a subcell is often defined in a barycentric space lower than that of the cell.

The next property is essential to later define topological constraints using a set of incidence constraints.

PROPERTY.– Let C_1 be a cell defined in BI^n by an IFS $\mathbb{T}^1 = \{T_i^1\}_{i \in \Sigma_1}$ and C_2 a cell defined in BI^m by an IFS $\mathbb{T}^2 = \{T_j^2\}_{j \in \Sigma_2}$ ($n \leq m$) and $\amalg$ an incidence operator of C_1 to C_2. If:

$$\forall i \in \Sigma_1, \exists j \in \Sigma_2 \ tq \ \amalg T_i^1 = T_j^2 \amalg \qquad [1.5]$$

then:

$$\amalg(\mathcal{A}(\mathbb{T}^1)) \subset \mathcal{A}(\mathbb{T}^2)$$

PROOF.– Let us denote: $\mathcal{A}_1 = \mathcal{A}(\mathbb{T}^1)$:

$$\amalg(A_1) = \amalg(\bigcup_{i \in \Sigma_1} T_i^1 A_1)$$

$$= \bigcup_{i \in \Sigma_1} (\amalg T_i^1 A_1)$$

$$= \bigcup_{j \in \Sigma_1 \subset \Sigma_2} T_j^2 (\amalg A_1)$$

Therefore, $\amalg(A_1) \subset BI^m$ is the fixed point of $\{T_j^2\}_{j \in \Sigma_1}$, that is to say, $\amalg(A_1) = \mathcal{A}(\{T_j^2\}_{j \in \Sigma_1})$. Since $\Sigma_1 \subset \Sigma_2$; $\amalg(A_1) \subset \mathcal{A}(\mathbb{T}^2)$.

DEFINITION.– *Equations [1.5] are called incidence equations between two IFS and define a constraint of incidence between attractors.*

This incidence constraint requires a subset of cell transformations to contract the space in a similar manner to the transformations of the subcell, though only for the dimensions mapped by the incidence operator. As a result, the sub-attractor of the cell corresponding to this subset of transformations will be identical to the subcell attractor (up to an embedding Π).

EXAMPLE.– Let us take a second example for the construction of incidence between a curve and a face. We are going to build a Sierpinski triangle, whose three edges are curves of the "Takagi" type (see Figure 1.43, left). We choose to define our face from four control points: P_0, P_1, P_2 and P_3. Its attractor must therefore be defined in a barycentric space of dimension 4 (that is, BI^4). We denote its IFS by $\mathbb{T}^F = \{T_0^F, T_1^F, T_2^F\}$. It consists of three transformations, represented by (4×4) matrices. In order to be able to visualize $\mathcal{A}^F = \mathcal{A}(\mathbb{T}^F)$, we calculate its projection according to the four control points (see Figure 1.43).

Let us consider the curve whose projection following the control points Q_0, Q_1, Q_2 is illustrated by Figure 1.42. This curve is defined in BI^3 (because it depends on three control points) and is divided into two parts. We denote its IFS by $\mathbb{T}^e = \{T_0^e, T_1^e\}$ and its attractor by $\mathcal{A}^e = \mathcal{A}(\mathbb{T}^e)$ (that is, the curve). In this example: $T_0^e = \begin{pmatrix} 1 & .2 & .4 \\ 0 & .8 & .2 \\ 0 & 0 & .4 \end{pmatrix}$

and $T_1^e = \begin{pmatrix} .4 & 0 & 0 \\ .2 & .8 & 0 \\ .4 & .2 & 1 \end{pmatrix}$.

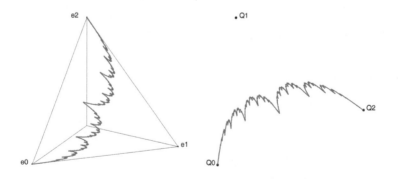

Figure 1.42. *Curve of the "Takagi" type, defined from three control points and two transformations. For a color version of this figure, see www.iste.co.uk/gentil/geometric.zip*

Using incidence constraints, we shall require that this curve corresponds to the edges of the face. The association between control points and the dimensions of the barycentric space is implicitly made by the indices of the control points. For the curve, Q_0 corresponds to the first dimension, Q_1 to the second and Q_2 to the third. Similarly, for the face, P_0 corresponds to the first, etc.

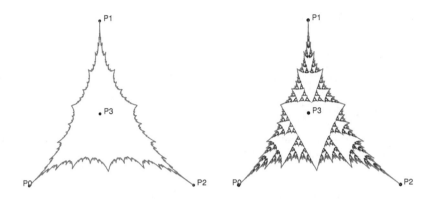

Figure 1.43. *Incidence constraints. On the left: Three curves of the "Takagi" type, defined as for Figure 1.42. On the right: Using the incidence equations, the curves are imposed as edges of the "face" of the Sierpinski triangle type. For a color version of this figure, see www.iste.co.uk/gentil/geometric.zip*

To obtain a Sierpinski triangle whose edges correspond to the curve shown in Figure 1.42, we shall first define the incidence operator for each face edge. We recall that this operator simply has to map the dimensions of the barycentric space of the curve with those of the barycentric space of the face. Thus, for the edge of the face going from vertex P_0 to vertex P_1, the incidence operator is: $\amalg_0 = \begin{pmatrix} 1 & 0 & 0 \\ 0 & 0 & 1 \\ 0 & 0 & 0 \\ 0 & 1 & 0 \end{pmatrix}$.

Similarly, for the edges joining vertex P_1 to vertex P_2, the incidence operator is $\amalg_1 = \begin{pmatrix} 0 & 0 & 0 \\ 1 & 0 & 0 \\ 0 & 0 & 1 \\ 0 & 1 & 0 \end{pmatrix}$ and, for the edge between P_2 and P_0, $\amalg_2 = \begin{pmatrix} 0 & 0 & 1 \\ 0 & 0 & 0 \\ 1 & 0 & 0 \\ 0 & 1 & 0 \end{pmatrix}$. These three curves will have P_3 as the control point associated with the second dimension (e_1).

Once the incidence operators are defined, we can write the incidence constraints:

$$T_0^F \amalg_0 = \amalg_0 T_0^e \qquad [1.6]$$

$$T_1^F \amalg_0 = \amalg_0 T_1^e \qquad [1.7]$$

to ensure that $\amalg_0(\mathcal{A}^e) \subset \mathcal{A}^F$,

$$T_1^F \amalg_1 = \amalg_1 T_0^e \qquad\qquad [1.8]$$

$$T_2^F \amalg_1 = \amalg_1 T_1^e \qquad\qquad [1.9]$$

to ensure that $\amalg_1(\mathcal{A}^e) \subset \mathcal{A}^F$, and finally:

$$T_2^F \amalg_2 = \amalg_2 T_0^e \qquad\qquad [1.10]$$

$$T_0^F \amalg_2 = \amalg_2 T_1^e \qquad\qquad [1.11]$$

to ensure that $\amalg_2(\mathcal{A}^e) \subset \mathcal{A}^F$.

By denoting the coefficient of the ith row and the jth column of T_0^F by t_{ij}, we deduce incidence equations [1.6] and [1.11] such that:

$$T_0^F \amalg_0 = \begin{pmatrix} t_{00} & t_{03} & t_{01} \\ t_{10} & t_{13} & t_{11} \\ t_{20} & t_{23} & t_{21} \\ t_{30} & t_{33} & t_{31} \end{pmatrix} = \amalg_0 T_0^e = \begin{pmatrix} 1 & .2 & .4 \\ 0 & 0 & .4 \\ 0 & 0 & 0 \\ 0 & .8 & .2 \end{pmatrix} \qquad [1.12]$$

$$T_0^F \amalg_2 = \begin{pmatrix} t_{02} & t_{03} & t_{00} \\ t_{12} & t_{13} & t_{10} \\ t_{22} & t_{23} & t_{20} \\ t_{32} & t_{33} & t_{30} \end{pmatrix} = \amalg_2 T_1^e = \begin{pmatrix} .4 & .2 & 1 \\ 0 & 0 & 0 \\ .4 & 0 & 0 \\ .2 & .8 & 0 \end{pmatrix} \qquad [1.13]$$

It finally yields: $T_0^F = \begin{pmatrix} 1 & .4 & .4 & .2 \\ 0 & .4 & 0 & 0 \\ 0 & 0 & .4 & 0 \\ 0 & .2 & .2 & .8 \end{pmatrix}$.

From the other equations, we deduce:

$$T_1^F = \begin{pmatrix} .4 & 0 & 0 & 0 \\ .4 & 1 & .4 & .2 \\ 0 & 0 & .4 & 0 \\ .2 & 0 & .2 & .8 \end{pmatrix}; T_2^F = \begin{pmatrix} .4 & 0 & 0 & 0 \\ 0 & .4 & 0 & 0 \\ .4 & .4 & 1 & .2 \\ .2 & .2 & 0 & .8 \end{pmatrix}$$

Similarly, using the pairs of equations, [1.7] and [1.8], then [1.9] and [1.10], we determine the coefficients of T_1^F and T_2^F, respectively.

REMARK.– In this example, incidence constraints completely determine the coefficients of the matrices of subdivision operators. This is not always the case. There may remain degrees of freedom with unfixed coefficient values. In this case, incidence relations will be retained, regardless of the values of these coefficients. Conversely, the constraints may be too strong and the system may not have a solution. It will then always be possible to increase the number of degrees of freedom of the system (by adding dimensions to barycentric spaces, for example, of the control points) to obtain a solution.

At this stage, we have access to tools (incidence operators and incidence constraints) to build a hierarchy of cells in a similar way to B-rep models, but with the difference that cells may be of any topological structure. On the other hand, we only have the guarantee of the inclusion of a subcell into a cell. We define a structuring of the space with a hierarchy of subspaces associated with cells. This hierarchy has an important consequence, which we will highlight later: it allows us to prioritize the influences of the control points.

Incidence operators can be interpreted from two points of view. The first presentation above presents incidence operators as operators embedding a subcell into a cell: $\amalg \mathcal{A}_n \subset \mathcal{A}_m$. The second point of view, as we have explained for subdivision transformations, consists of considering the left multiplication of the incidence operator by control points: $[P_0 \cdots P_{m-1}]\amalg$. This amounts to extracting (or selecting) a subset of control points, which will be used to project the subcell. For example, for the previous example:

$$[P_0 P_1 P_2 P_3]\amalg_0 = [P_0 P_1 P_2 P_3] \begin{pmatrix} 1 & 0 & 0 \\ 0 & 0 & 1 \\ 0 & 0 & 0 \\ 0 & 1 & 0 \end{pmatrix} = [P_0 P_3 P_1]$$

Therefore, P_0, P_3 and P_1 will be the three control points of the first curve incident to the face.

The incidence constraints constitute the first step toward defining the cellular decomposition of the attractor. The second step consists of using this cellular decomposition, in the subdivision process, to constraint the attractor by defining "connections" between copies of itself and thus to define a topological structure.

Let us take the example of Figure 1.44. We chose an IFS consisting of four transformations: $\mathbb{T} = \{T_0, T_1, T_2, T_3\}$ whose attractor is defined within a barycentric space BI^8. At first, we imposed curves (of the "Takagi curve" type) incident to the attractor. These curves are shown in black in Figure 1.44(a). In this same figure, no connection constraints (adjacency relation) have been defined between the subdivisions of the attractor. Therefore, the four subdivisions (in red, green, purple and blue) connect with each other two by two, following a single point, due only to the incidence constraints. For Figure 1.44(b), a connection has been defined between the subdivision shown in red ($T_0(\mathcal{A})$) and the one shown in green ($T_1(\mathcal{A})$), via their respective right ($\amalg_1$) and left ($\amalg_3$) edges (in orange). This constraint is simply written in the form: $T_0\amalg_1 = T_1\amalg_3$. The resolution of this constraint results in the equality of certain columns (up to a permutation) of matrices T_0 and T_1.

Figure 1.44(b) shows the result of applying this constraint while maintaining the coefficients of the matrix T_0. Figure 1.44(c) shows the result of the application of a connection between the subdivisions represented in green and purple, by the constraint: $T_1 \amalg_2 = T_2 \amalg_0$ (having kept the coefficients of T_1). Similarly, a connection has been defined between the subdivisions depicted in purple and blue, to obtain Figure 1.44(d). Finally, for Figure 1.44(e) all the connections have been made, thus defining the topology of a surface.

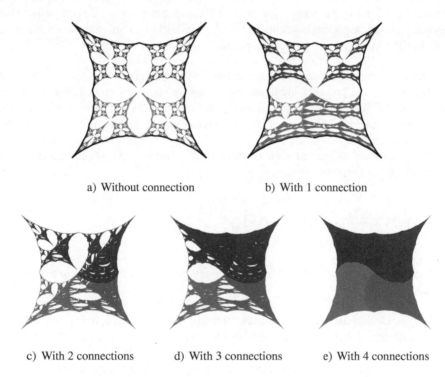

a) Without connection   b) With 1 connection

c) With 2 connections   d) With 3 connections   e) With 4 connections

Figure 1.44. *Example of the construction of a connection between the subdivisions of an attractor. For a color version of this figure, see www.iste.co.uk/gentil/geometric.zip*

COMMENTS ON FIGURE 1.44.– *(a) The attractor $\mathcal{A}$ is represented with four incident curves (of the "Takagi" type, in black) defining the edges. The subdivisions of this attractor are as follows: $T_0(\mathcal{A})$ in red, $T_1(\mathcal{A})$ in green, $T_2(\mathcal{A})$ in purple and $T_3(\mathcal{A})$ in blue. These different parts are disjoint, except along the edges where they connect, two by two, by a point. The edges on the right-hand side of the red subdivision ($T_0 \amalg_1$) and on the left-hand side of the green subdivision ($T_1 \amalg_3$) are represented in orange. (b) An adjacency relation has been defined between the red subdivision ($T_0(\mathcal{A})$) and* *the green subdivision $T_1(\mathcal{A})$, through their orange edges, by the constraint: $T_0 \amalg_1 = T_1 \amalg_3$. This constraint implies the sharing of columns of T_0 and T_1. The coefficients of T_0 have been preserved, inducing a deformation of the subdivision $T_1(\mathcal{A})$ (in green). (c) An adjacency relation has been added to (b), connecting the green subdivision and the purple subdivision: $T_1 \amalg_2 = T_2 \amalg_0$. (d) A third adjacency relation connects the purple and blue parts. (e) A fourth adjacency relation makes it possible to obtain a surface.*

To clarify and better understand this topological construction, let us analyze the definition of connections between the parts of the attractor more formally. To do this, consider an IFS $\mathbb{T} = \{T_i\}_{i \in \Sigma}$ defined in a barycentric space BI^n and denote its attractor by $\mathcal{A}$. Suppose we have defined a set of cells on $\mathcal{A}$, denoted $\mathcal{A}_k, k \in \Sigma'$, with $\Sigma' \subset \Sigma$. Let us denote the incidence operator of each cell by reversed PI_k. Given our concept of incident cell as a simple inclusion, the transformation $T(\mathcal{A})$ of $\mathcal{A}$ by an arbitrary transformation T (even non-continuous) has the same cell and sub-cell structure as $\mathcal{A}$ (because if a set B is included in another set C, then $T(B) \subset T(C)$). This is the case, *a fortiori*, of transformations T_i composing the IFS. Thus, each subdivision $T_i(\mathcal{A}_k)$ of the attractor has the same cell structure as $\mathcal{A}$ itself, that is, $T_i(\mathcal{A}_k) \subset T_i(\mathcal{A}), k \in \Sigma'$. We shall use this structure to make connections between subdivisions of $\mathcal{A}$ by sharing subcells. This constraint is defined by adjacency constraints of the form $T_j \amalg_l \mathcal{A}_l = T_j \amalg_m \mathcal{A}_m$ with $i, j \in \Sigma$ and $l, m \in \Sigma'$. The interpretation of this constraint is simple and intuitive:

– $\amalg_l \mathcal{A}_l$ represents the cell l of $\mathcal{A}$ embedded in $\mathcal{A}$ by $\amalg_l$;

– $\amalg_m \mathcal{A}_m$ represents the cell m of $\mathcal{A}$ embedded in $\mathcal{A}$ by $\amalg_m$;

– $T_i \amalg_l \mathcal{A}_l$ is the cell l of the subdivision $T_i \mathcal{A}$;

– analogously, $T_j \amalg_m \mathcal{A}_m$ is the cell m of the subdivision j of $\mathcal{A}$;

– the adjacency constraint requires that these two sets be the same.

This implies, on the one hand, that the spaces in which the cells are defined are the same and, on the other hand, that the cells themselves are identical ($\mathcal{A}_l = \mathcal{A}_m$), thus, defined, by the same sub-IFS. However, we want this constraint to be verified independently of the attractor by which the connection is made (that is, $\mathcal{A}_l = \mathcal{A}_m$) in order to be able to modify the parameters (the IFS) defining $\mathcal{A}_l$ while maintaining the connection between the subdivisions of $\mathcal{A}$. To this end, the constraint must be stronger and becomes:

$$T_i \amalg_l = T_k \amalg_m$$

1.3.6. *Formulation*

The formulation of BC-IFS simply consists of adding incidence and adjacency constraints to the C-IFS, or more precisely, defining the equivalency relations between the words accepted by the automaton.

A BC-IFS is given by the following elements.

1) An automaton $(\Sigma, Q, \delta, \natural)$ where:

- Σ is an alphabet partitioned into: a set of subdivision symbols, denoted by $\Sigma_{\div} = \{\div_0, \div_1, \dots\}$; a set of incidence symbols, denoted by $\Sigma_{\partial} = \{\partial_0, \partial_1, \dots\}$;

- Q represents a set of terminal states with for every state $q \in Q$: an associated space $(\mathbb{X}^q)_{q \in Q}$; a display primitive $(K^q)_{q \in Q}$ where $K^q, \in \mathcal{H}(\mathbb{X}^q)$;

- δ is a transition function $\delta : Q \times \Sigma \rightarrow Q$;

- $\natural$ is the initial state.

To make the text easier to read, we will indicate the starting state of the transition by using the superscript of the symbol to which it corresponds. Therefore, for $q \in Q$ we shall denote:
- the outbound subdivisions $\div_i^q$;

- the outbound incidences ∂_i^q.

The transition function δ is extended onto the set of finite words Σ^* according to this recursive definition:

- $\forall q \in Q, \delta(q, \varepsilon) = q$ where ε is the empty word;

- $\forall (q, \sigma, \theta) \in (Q \times \Sigma \times \Sigma^*), \delta(q, \sigma\theta) = \delta(\delta(q, \sigma), \theta)$.

2) A set of operators: With every symbol $\sigma \in \Sigma^q = \{\tau | (q, \tau) \in \mathcal{D}_\delta\}$, namely, the set of the outgoing transitions of state q, we associate:

- a subdivision operator if $\sigma \in \Sigma_{\div}^q$:

$$T_\sigma^q : E^{\delta(q,\sigma)} \rightarrow E^q$$

- an incidence operator if $\sigma \in \Sigma_{\partial}^q$:

$$\amalg_\sigma^q : E^{\delta(q,\sigma)} \rightarrow E^q$$

The notation $\hat{T}_i^q$ indiscriminately designates the subdivision or incidence operator, that is: $\hat{T}_i^q = T_i^q$ if $i \in \Sigma_{\div}^q$ and $\hat{T}_i^q = \amalg_i^q$ if $i \in \Sigma_{\partial}^q$. This notation will allow us to generically write the equivalence rules.

Apart from the distinction made between the symbols and the associated operators, the automaton of a BC-IFS is identical to that of a C-IFS. It generates a set of words $L(\natural, Q)$. These words represent expressions describing figures. It has the same characteristics, including:

- with every state $q \in Q$ a figure $\mathcal{A}^q = \bigcup_{\sigma \in \Sigma_q} \hat{T}_\sigma^q \mathcal{A}^{\delta(q,\sigma)}$ is associated;

- a path of the automaton (represented by a finite word θ of length n, $\theta \in L^n(q, Q) = \Sigma^n \cap \mathcal{D}_\delta$) represents an expression describing a component of $\mathcal{A}$ defined by $\hat{T}_\theta^q \mathcal{A}^{\delta(q,\theta)}$;

- each component of a finite path of label $\theta \in L^n(q, Q)$ is approximated by a figure $K_\theta^q = \hat{T}_\theta^q K^{\delta(q,\theta)}$, which is effectively computed.

3) A set of equivalence relations and rules of incidence and adjacency.

The equivalence rules are the base of the BC-IFS model. They are used to define incidence and adjacency relations.

DEFINITION.– *Two paths γ_σ^q and $\gamma_{\sigma'}^q$ are said to be equivalent if:*

$$\gamma_\sigma^q \simeq \gamma_{\sigma'}^q \overset{def}{\Leftrightarrow} \delta(q, \sigma) = \delta(q, \sigma') \ et \ \hat{T}_\sigma^q = \hat{T}_{\sigma'}^q$$

PROPERTY.– Two equivalent paths describe the same component of the attractor and consequently, the same approximation figures.

PROOF.– Given that the arrival states are the same ($\partial(\natural, \sigma) = \partial(\natural, \sigma')$), we have: $\mathcal{A}^{\partial(\natural,\sigma)} = \mathcal{A}^{\partial(\natural,\sigma')}$ and $K^{\partial(\natural,\sigma)} = K^{\partial(\natural,\sigma')}$. In addition, since $\hat{T}_\sigma^\natural = \hat{T}_{\sigma'}^\natural$ then $\hat{T}_\sigma^\natural \mathcal{A}^{\partial(\natural,\sigma)} = \hat{T}_{\sigma'}^\natural \mathcal{A}^{\partial(\natural,\sigma')}$ and $\hat{T}_\sigma^\natural K^{\partial(\natural,\sigma)} = \hat{T}_{\sigma'}^\natural K^{\partial(\natural,\sigma')}$, that is, $K_\sigma^\natural = K_{\sigma'}^\natural$.

From this definition, we introduce the set of equivalence rules, denoted by Υ^q, containing the pairs of equivalent paths (σ, σ'):

$$(\sigma, \sigma') \in \Upsilon^q \Leftrightarrow \begin{cases} \partial(q, \sigma) = \partial(q, \sigma') \\ \hat{T}_\sigma^q = \hat{T}_{\sigma'}^q \end{cases}$$

From this equivalence relation, we define three types of rules:

– incidence equations:

$$(\partial_i \div_j, \div_k \partial_l) \in \Upsilon^q$$

that describe that a sub-cell belongs to a cell:

$$(\div_i \partial_j, \div_k \partial_l) \in \Upsilon^q$$

– adjacency equations: to connect two subdivisions via a subcell;

– adjacency equations on incidence operators:

$$(\partial_i \partial_j, \partial_k \partial_l) \in \Upsilon^q$$

ensuring the consistency of incidence operators (see note from section 1.3.2).

1.3.7. *Evaluation tree and quotient graph*

The evaluation tree, defined for a C-IFS and still valid for a BC-IFS, creates a topological structure equivalent to that of a Cantor set. The addition of the equivalency rules will change this topological structure, which will then be represented by the quotient tree.

It should be recalled that in order to define a figure, we need to specify the following elements:

– the cell decomposition specifying which subcells are incident to which cells;

– for every cell:

  - how many parts it subdivides into (this induces the number of subdivision operators),

  - the size of the barycentric space, that is, on how many control points does the cell depend (incidence operators will specify the distribution of control points between the different cells),

  - a display primitive;

– incidence and adjacency constraints.

Let us illustrate these elements for the construction of a curve. For our example, we want our curve to depend on three control points, each vertex to depend on a single control point and to be able to divide the curve into two parts. An example of this type of curve is given in Figure 1.42.

– The cellular decomposition of a curve is always the same: the curve represents the highest level cell and it is bordered by two vertices (subcells).

– Since the curve can be divided into two parts, we will have two subdivision operators.

– It must be defined in a barycentric space of dimension 3, since it depends on three control points.

– Each vertex only depends on a single control point, each is defined within a barycentric space of dimension 1 (BI^1). The vertices still only have one subdivision operator. In the trivial case of BI^1, this operator associates the unique point of BI^1n, that is, 1, to itself (see section 1.3.5).

The automaton in Figure 1.45 represents this cellular decomposition and the associated subdivision process.

Figure 1.46 represents the evaluation tree produced from the automaton in Figure 1.45 for words of length 2. To differentiate the symbols, the tree is represented by subdivision levels: the first line corresponds to the first level, and the second line corresponds to the second level. The actions of the symbols of incidence operators are represented horizontally, while the actions of the symbols of subdivision operators are represented from the first level to the second level.

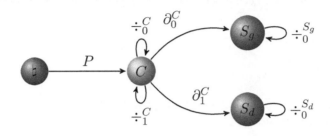

Figure 1.45. *Automaton integrating the cellular decomposition of a curve subdividing into two parts. For a color version of this figure, see www.iste.co.uk/gentil/geometric.zip*

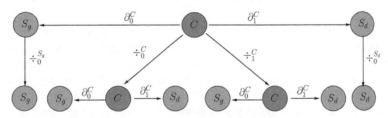

Figure 1.46. *Tree for a curve. For a color version of this figure, see www.iste.co.uk/gentil/geometric.zip*

This tree represents what we obtain if we choose arbitrary subdivision operators $(T_i^{\div})$, that is a set of cells disconnected from one another.

Now we need to clarify the incidence operators:

– if we want the left vertex (S_g) to depend on the first control point, the first incidence operator must be: $\amalg_0^C = \begin{pmatrix} 1 \\ 0 \\ 0 \end{pmatrix}$;

· – similarly, for the right vertex to depend on the third control point, the second incidence operator must be: $\amalg_1^C = \begin{pmatrix} 0 \\ 0 \\ 1 \end{pmatrix}$.

This choice is arbitrary and depends on the role that we wish to assign to each control point. This first step maps the dimensions of the barycentric subspaces of the vertices with the dimensions of the barycentric space of the curve, but does not guarantee the incidence of the vertices to the curve. To ensure that vertices are incidental to the curve, we need to write down the following incidence constraints:

$$\partial_0^C \div_0^{S_g} \simeq \div_0^C \partial_0^C \qquad\qquad [1.14]$$

$$\partial_1^C \div_0^{S_d} \simeq \div_1^C \partial_1^C \qquad\qquad [1.15]$$

Finally, the connection between the two subdivisions of the curve is defined by the following adjacency constraint:

$$\div_0^C \partial_1^C \simeq \div_1^C \partial_0^C \qquad\qquad [1.16]$$

By definition of the equivalence relation, this adjacency constraint implies equality between the arrival states $S_g = S_d$, which we denote by S. We denote its subdivision operator by $\div_0^S$. This constraint is intuitive, in the sense that if we want to connect the subdivisions of the curve by the right and left vertices, these vertices must be constituted by the same attractor. From the equivalence relation between the paths of the tree, we can build the quotient graph symbolizing the resulting topological structure (see Figure 1.47). We can observe that at each level of iteration, we have a topological structure equivalent to that of a curve. For level 1, the curve is bordered by two vertices. At level 2, the curve is composed of two parts of the curve, each bordered by two vertices, connected by means of a vertex.

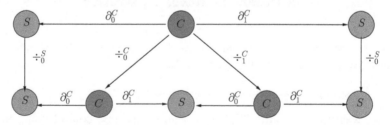

Figure 1.47. *Quotient graph for a curve. For a color version of this figure, see www.iste.co.uk/gentil/geometric.zip*

1.3.8. *Constraint resolution*

The topological structure symbolized by the quotient graph [1.47] is geometrically reflected when subdivision operators are instantiated (the incidence operators have already been set). This instantiation must satisfy the equations induced on the operators by the equivalence relations. For the example of the curve: the incidence relations [1.14] and [1.15] and the adjacency relation [1.16] are reflected in the operators by the equations:

$$\amalg_0^C T_0^S = T_0^C \amalg_0^C \tag{1.17}$$

$$\amalg_1^C T_0^S = T_1^C \amalg_1^C \tag{1.18}$$

$$T_0^C \amalg_1^C = T_1^C \amalg_0^C \tag{1.19}$$

with:

$$ - \amalg_0^C = \begin{pmatrix} 1 \\ 0 \\ 0 \end{pmatrix} ; \amalg_1^C = \begin{pmatrix} 0 \\ 0 \\ 1 \end{pmatrix} ; $$

$- T^S = 1$ (the only subdivision operator possible in BI^1);

$- T_0^C$, is a (3×3) matrix;

$- T_1^C$ is a (3×3) matrix.

Let us name the coefficients of subdivision matrices: $T_0^C = \begin{pmatrix} a & d & g \\ b & e & h \\ c & f & i \end{pmatrix}$ and

$T_1^C = \begin{pmatrix} a' & d' & g' \\ b' & e' & h' \\ c' & f' & i' \end{pmatrix}$.

The resolution of incidence and adjacency constraints is simple, since on each side of each equation we have the product of an incidence matrix with a subdivision matrix. Due to the nature of the incidence matrix (permutation matrix), these products correspond to a column extraction from the subdivision matrix:

$$\amalg_0^C T_0^S = T_0^C \amalg_0^C$$

$$\begin{pmatrix} 1 \\ 0 \\ 0 \end{pmatrix} \times (1) = \begin{pmatrix} a & d & g \\ b & e & h \\ c & f & i \end{pmatrix} \begin{pmatrix} 1 \\ 0 \\ 0 \end{pmatrix}$$

$$\begin{pmatrix} 1 \\ 0 \\ 0 \end{pmatrix} = \begin{pmatrix} a \\ b \\ c \end{pmatrix}$$

The solution of equation [1.17] requires that the first column of the operator T_0^C contains the value 1 (that is, T_0^S) with zeroes underneath, in the first row. We shall see the nature of this constraint later.

$$\amalg_1^C T_0^S = T_1^C \, \amalg_1^C$$

$$\begin{pmatrix} 0 \\ 0 \\ 1 \end{pmatrix} \times 1 = \begin{pmatrix} a' & d' & g' \\ b' & e' & h' \\ c' & f' & i' \end{pmatrix} \begin{pmatrix} 0 \\ 0 \\ 1 \end{pmatrix}$$

$$\begin{pmatrix} 0 \\ 0 \\ 1 \end{pmatrix} = \begin{pmatrix} g' \\ h' \\ i' \end{pmatrix}$$

The solution of equation [1.18] requires that the last column of the operator T_1^C contains the value 1 (that is, T_0^S) with zeroes underneath, in the last row.

$$T_0^C \amalg_1^C = T_1^C \, \amalg_0^C$$

$$\begin{pmatrix} a & d & g \\ b & e & h \\ c & f & i \end{pmatrix} \begin{pmatrix} 0 \\ 0 \\ 1 \end{pmatrix} = \begin{pmatrix} a' & d' & g' \\ b' & e' & h' \\ c' & f' & i' \end{pmatrix} \begin{pmatrix} 1 \\ 0 \\ 0 \end{pmatrix}$$

$$\begin{pmatrix} g \\ h \\ i \end{pmatrix} = \begin{pmatrix} a' \\ b' \\ c' \end{pmatrix}$$

The solution of equation [1.19] requires that the last column of the operator T_0^C corresponds to the first column of the operator T_1^C.

Thereby, to define a curve:

– from two subdivision operators;

– we need three control points;

– whose vertices each depend on a single control point, the structure of the subdivision operators must be:

$$T_0^C = \begin{pmatrix} 1 & \bullet & x \\ 0 & \bullet & y \\ 0 & \bullet & z \end{pmatrix} \; ; T_1^C = \begin{pmatrix} x & \bullet & 0 \\ y & \bullet & 0 \\ z & \bullet & 1 \end{pmatrix}$$

where $x, y, z \in \mathbb{R}$ and the symbol $\bullet$ indicate the location of any real coefficient. The parameters can be arbitrarily chosen, provided that:

- the sum of the coefficients of each column must be equal to 1,

- the operator must be contractive.

This last constraint can be verified using eigenvalues: the second largest eigenvalue must be of modulus < 1 (the largest is equal to 1 and always exists for this type of matrix). We find that the operators subdividing the curve in Figure 1.42 verify all of these conditions. Three other examples are presented in Figure 1.48.

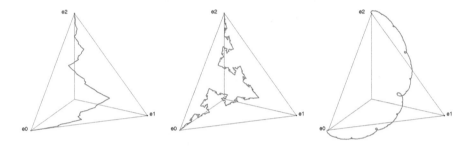

Figure 1.48. *Example of curves generated for different parameter values. For a color version of this figure, see www.iste.co.uk/gentil/geometric.zip*

A special case is that of De Casteljau matrices generating Bernstein polynomials of the second degree: $B_0^C = \begin{pmatrix} 1 & \frac{1}{2} & \frac{1}{4} \\ 0 & \frac{1}{2} & \frac{1}{2} \\ 0 & 0 & \frac{1}{4} \end{pmatrix}$ and $B_1^C = \begin{pmatrix} \frac{1}{4} & 0 & 0 \\ \frac{1}{2} & \frac{1}{2} & 0 \\ \frac{1}{4} & \frac{1}{2} & 1 \end{pmatrix}$ (see Figure 1.49).

The resulting attractor corresponds to a curve because its IFS responds to the incidence and adjacency constraints of a curve. In this case, the curve corresponds to the second-degree Bernstein polynomial.

To fully illustrate the equivalence relation, let us consider the more complete example of a surface tile subdividing into four (see Figure 1.50). We have introduced the concept of adjacency relations between incidence operators. This type of relation was not presented for the example of the curve, because the topological structure is not rich enough to require it (two levels of cells). In the case of a surface, we identify face-type cells. These are bordered by edge-type cells, themselves bordered by vertex-type cells.

The automaton of such a subdivision system is represented by Figure 1.51. In the definition of the automaton, we have anticipated the consequences of adjacency constraints, only specifying two types of edges. Indeed, similarly to the vertex for the curve, the connection of two face subdivisions by the edges requires the states of the corresponding edges to be identical. In this example, the connections are made along the "horizontal" and "vertical" edges. Therefore, we do not need more than two types of edges. Similarly, we could have defined a much larger number of vertex types, but equivalence relations dictate that there is only one type of vertex. We could not make any it *a priori* assumption and thus define as many cells as necessary. The equivalence relations would have induced the equalities between the cells. To simplify the representation of the automaton, we have anticipated these constraints.

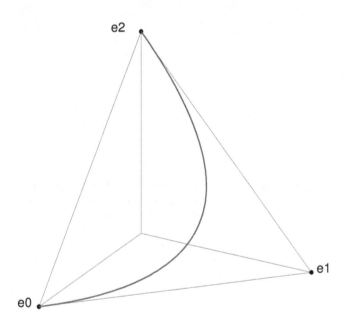

Figure 1.49. *Attractor built from the IFS* $\{B_0^C, B_1^C\}$ *whose subdivision operators correspond to De Casteljau matrices* $B_0^C = \begin{pmatrix} 1 & \frac{1}{2} & \frac{1}{4} \\ 0 & \frac{1}{2} & \frac{1}{2} \\ 0 & 0 & \frac{1}{4} \end{pmatrix}$ *and* $B_1^C = \begin{pmatrix} \frac{1}{4} & 0 & 0 \\ \frac{1}{2} & \frac{1}{2} & 0 \\ \frac{1}{4} & \frac{1}{2} & 1 \end{pmatrix}$.

For a color version of this figure, see www.iste.co.uk/ gentil/geometric.zip

Figure 1.52 represents the tree corresponding to the subdivision of a quadrangular surface, described by the automaton in Figure 1.51. This diagram represents what we obtain without incidence and adjacency constraints. The face F, edges a and b and vertex S states have attractors (topological cells) separated from one another. If we want to obtain a continuous surface, we have to write the constraints on the edges (states a and b), so that the attractors are actually curves. That is, for the cell a:

$$\div_0^a \partial_1^a \simeq \div_1^a \partial_0^a$$
$$\partial_0^a \div_0^S \simeq \div_0^a \partial_0^a$$
$$\partial_1^a \div_0^S \simeq \div_1^a \partial_1^a$$

and for the cell b:

$$\div_0^b \partial_1^b \simeq \div_1^b \partial_0^b$$
$$\partial_0^b \div_0^S \simeq \div_0^b \partial_0^b$$
$$\partial_1^b \div_0^S \simeq \div_1^b \partial_1^b$$

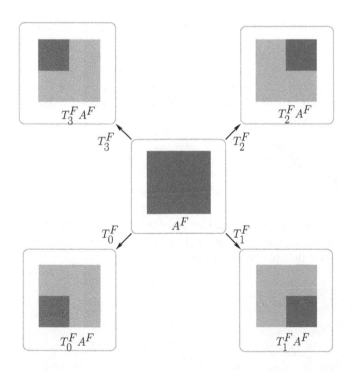

Figure 1.50. *Subdivision structure of the tile. For a color version of this figure, see www.iste.co.uk/gentil/geometric.zip*

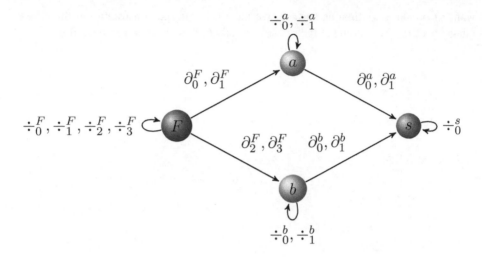

Figure 1.51. *Automaton for the subdivision of a quadrangular surface.*
For a color version of this figure, see www.iste.co.uk/gentil/geometric.zip

However, at this level, there is no guarantee that the edges share the same vertices. To this end, we need to write down the adjacency constraints on the incidence operators. There are four of them, one per face vertex (see the red dotted circle in Figure 1.52):

$$\begin{cases} \partial_0^F \partial_1^a \simeq \partial_1^F \partial_0^b \\ \partial_1^F \partial_1^b \simeq \partial_2^F \partial_0^a \\ \partial_2^F \partial_1^a \simeq \partial_3^F \partial_0^b \\ \partial_3^F \partial_0^b \simeq \partial_0^F \partial_0^a \end{cases}$$

Then, we need to specify the incidence constraints, to ensure that the two type-a curves and the two type-b curves are incidental to the face. There are eight of them, two per curve (see the blue dotted lines in Figure 1.52):

$$\begin{cases} \div_0^F \partial_0^F \simeq \partial_0^F \div_0^a \\ \div_1^F \partial_0^F \simeq \partial_0^F \div_1^a \end{cases}$$

$$\begin{cases} \div_1^F \partial_1^F \simeq \partial_0^F \div_0^b \\ \div_2^F \partial_1^F \simeq \partial_0^F \div_1^b \end{cases}$$

$$\begin{cases} \div_2^F \partial_2^F \simeq \partial_2^F \div_1^a \\ \div_3^F \partial_2^F \simeq \partial_2^F \div_0^a \end{cases}$$

$$\begin{cases} \div_3^F \partial_3^F \simeq \partial_3^F \div_1^b \\ \div_0^F \partial_3^F \simeq \partial_0^F \div_0^b \end{cases}$$

All that remains is to write down the adjacency relations, so that the subdivisions of the faces are connected to each other. There are four of them (see the red dotted lines in Figure 1.52):

$$\begin{cases} \div_0^F \partial_1^F \simeq \div_1^F \partial_3^F \\ \div_3^F \partial_1^F \simeq \div_2^F \partial_3^F \\ \div_0^F \partial_2^F \simeq \div_3^F \partial_0^F \\ \div_1^F \partial_2^F \simeq \div_2^F \partial_0^F \end{cases}$$

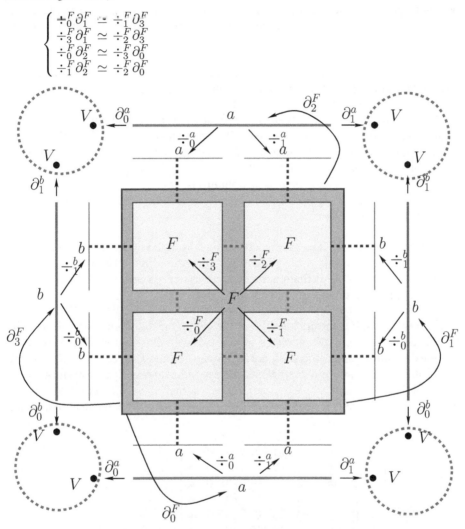

Figure 1.52. *Cell structure of a quadrangular tile. For a color version of this figure, see www.iste.co.uk/gentil/geometric.zip*

REMARK.– In the presentation of this example we did not specify the dimensions of the barycentric spaces for each cell. Indeed, this is not necessary. The topological structure is defined independently of any geometric consideration. This means that the incidence relations that we have described are valid and can be reused regardless of the geometric context: the number of control points assigned to each cell. Naturally, there is consistency between the sets of control points: one face integrates the control points of the edges that border it, and two adjacent edges share the same control points. Nonetheless, these consistencies are automatically determined by equivalence relations.

Let us choose a geometric context: Each vertex depends on a control point, each edge of three control points and the face of nine control points.

The solutions of the constraints are expressed by constraints between the subdivision matrices, in the form of equality between coefficients. The coefficients ax_i (bx_i) correspond to the coefficients of the curves bordering the face. Adjacency constraints are expressed by identical columns (up to a permutation) between subdivision matrices. For example, a column is common to the four matrices (coefficient ra_i). It corresponds to the central connection. Finally, for each matrix, there is a free column.

The presence of zero value coefficients comes from the incidence constraints. An incidence operator embeds a subcell (definite attractor within a barycentric space of dimension p) into a cell (definite attractor within a barycentric space of dimension $q > p$), such that the subcell is a sub-attractor of the cell.

In other words, if we restrict the IFS of the cell to the transformations involved in this incidence relation, the attractor associated with this restricted IFS remains contained within a barycentric subspace (of dimension p) of the cell space. For these same columns, the non-zero coefficients correspond to those of the transformations of the IFS of the subcell. We again, come across these properties for every interweaving of subcells incident to a cell, that is, for each vertex incident to a curve, for each curve incident to a face, etc.:

$$T_0^F = \begin{pmatrix} 1 & aa1 & ab1 & ra1 & rb1 & bb1 & rc1 & bc1 & \bullet \\ 0 & aa2 & ab2 & ra2 & rb2 & 0 & rc2 & 0 & \bullet \\ 0 & aa3 & ab3 & ra3 & rb3 & 0 & rc3 & 0 & \bullet \\ 0 & 0 & 0 & ra4 & rb4 & 0 & rc4 & 0 & \bullet \\ 0 & 0 & 0 & ra5 & rb5 & 0 & rc5 & 0 & \bullet \\ 0 & 0 & 0 & ra6 & rb6 & bb2 & rc6 & bc2 & \bullet \\ 0 & 0 & 0 & ra7 & rb7 & 0 & rc7 & 0 & \bullet \\ 0 & 0 & 0 & ra8 & rb8 & bb3 & rc8 & bc3 & \bullet \\ 0 & 0 & 0 & ra9 & rb9 & 0 & rc9 & 0 & \bullet \end{pmatrix}$$

$$
T_1^F = \begin{pmatrix}
aa1 & 0 & ac1 & 0 & 0 & ra1 & rd1 & rb1 & \bullet \\
aa2 & 1 & ac2 & bb1 & bc1 & ra2 & rd2 & rb2 & \bullet \\
aa3 & 0 & ac3 & 0 & 0 & ra3 & rd3 & rb3 & \bullet \\
0 & 0 & 0 & bb2 & bc2 & ra4 & rd4 & rb4 & \bullet \\
0 & 0 & 0 & bb3 & bc3 & ra5 & rd5 & rb5 & \bullet \\
0 & 0 & 0 & 0 & 0 & ra6 & rd6 & rb6 & \bullet \\
0 & 0 & 0 & 0 & 0 & ra7 & rd7 & rb7 & \bullet \\
0 & 0 & 0 & 0 & 0 & ra8 & rd8 & rb8 & \bullet \\
0 & 0 & 0 & 0 & 0 & ra9 & rd9 & rb9 & \bullet
\end{pmatrix}
$$

$$
T_2^F = \begin{pmatrix}
bb1 & ra1 & rc1 & 0 & rf1 & 0 & 0 & ba1 & \bullet \\
0 & ra2 & rc2 & 0 & rf2 & 0 & 0 & 0 & \bullet \\
0 & ra3 & rc3 & 0 & rf3 & 0 & 0 & 0 & \bullet \\
0 & ra4 & rc4 & aa2 & rf4 & 0 & ab2 & 0 & \bullet \\
0 & ra5 & rc5 & 0 & rf5 & 0 & 0 & 0 & \bullet \\
bb2 & ra6 & rc6 & aa1 & rf6 & 1 & ab1 & ba2 & \bullet \\
0 & ra7 & rc7 & aa3 & rf7 & 0 & ab3 & 0 & \bullet \\
bb3 & ra8 & rc8 & 0 & rf8 & 0 & 0 & ba3 & \bullet \\
0 & ra9 & rc9 & 0 & rf9 & 0 & 0 & 0 & \bullet
\end{pmatrix}
$$

$$
T_3^F = \begin{pmatrix}
ra1 & 0 & rd1 & 0 & 0 & 0 & 0 & rf1 & \bullet \\
ra2 & bb1 & rd2 & 0 & ba1 & 0 & 0 & rf2 & \bullet \\
ra3 & 0 & rd3 & 0 & 0 & 0 & 0 & rf3 & \bullet \\
ra4 & bb2 & rd4 & 1 & ba2 & aa2 & ac2 & rf4 & \bullet \\
ra5 & bb3 & rd5 & 0 & ba3 & 0 & 0 & rf5 & \bullet \\
ra6 & 0 & rd6 & 0 & 0 & aa1 & ac1 & rf6 & \bullet \\
ra7 & 0 & rd7 & 0 & 0 & aa3 & ac3 & rf7 & \bullet \\
ra8 & 0 & rd8 & 0 & 0 & 0 & 0 & rf8 & \bullet \\
ra9 & 0 & rd9 & 0 & 0 & 0 & 0 & rf9 & \bullet
\end{pmatrix}
$$

Figure 1.53 shows an example of a surface obtained for randomly selected coefficients.

Figure 1.53. *Example of a quadrangular surface. For a color version of this figure, see www.iste.co.uk/gentil/geometric.zip*

Figure 1.54 shows an instantiation of the coefficients needed to obtain Bezier curves on the edges of the surface. For this purpose, we use the De Casteljau matrix coefficients (see detail in Chapter 3). We observe that the resulting surface is not a tensor product of Bezier curves.

Figure 1.54. *Example of a quadrangular surface bordered by Bezier curves with an adjacency constraint removed: on the left with six iterations, and on the right with seven iterations. For a color version of this figure, see www.iste.co.uk/gentil/ geometric.zip*

Removing an adjacency constraint will introduce a hole in the surface that, by self-similarity, will appear at every iteration level (see Figure 1.55), and we define another topology. In terms of matrix structures, canceling an incidence constraint results in an increase in the number of degrees of freedom.

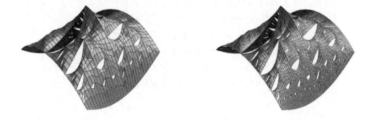

Figure 1.55. *Example of a quadrangular surface structure bordered by Bezier curves: on the left with six iterations, and on the right with seven iterations. For a color version of this figure, see www.iste.co.uk/gentil/geometric.zip*

The example in Figure 1.56 presents a structure with fractal topology, bordered by curves with fractal geometry. This example was taken from the automaton in Figure 1.28, to which we have added incidence and adjacency constraints. It should be noted that the face is defined by two states.

This approach can be applied to more sophisticated iterated systems such as those presented in Chapter 2.

Figure 1.56. *Example of a surface structure with fractal topology, obtained from the automaton in Figure 1.28: on the left with six iterations, and on the right with seven iterations. For a color version of this figure, see www.iste.co.uk/gentil/geometric.zip*

1.3.9. *Implementation*

The implementation of the BC-IFS model must take two distinct aspects into account. The first concerns the evaluation of the automaton. This evaluation is identical to that of the C-IFS. However, for the construction of the attractor, it is necessary to only apply subdivision operators to the highest level cell (the one that contains the others). Incidence operators are only present to define the incidence constraints of subcells in cells. Subcell attractors are automatically included in the attractors of the cells to which they are incident. It is not necessary to evaluate them. We shall therefore only go through the automaton by following the transitions corresponding to a subdivision symbol. Other evaluation algorithms can be developed. For example, we may choose to only visualize the curves bordering the faces. In this case, we will make the subdivisions of the faces up to the desired level of iteration and we will apply the transitions related to the incidence symbols to the curves. The latter can, in turn, be subdivided to reach the desired approximation. We will then be able to display the result of the primitive of the curve transformed by matrix products, corresponding to the words obtained.

The second aspect of the implementation is related to the resolution of equivalence relations defining the topology of the structure. As we have seen with the example of the curves, this resolution is almost trivial. The equations induced by the equivalence relation are composed on each side of the product equality of a matrix corresponding to a subdivision operator and a matrix corresponding to an incidence operator. The matrices of the incidence operators are permutation matrices supplemented by zeros. If the incidence operator is to the right of the matrix product, the product is equivalent to extracting columns from the second matrix by reorganizing them. If it is on the left, the product is equivalent to reorganizing the columns of the second matrix and inserting rows of zeros. Finally, the equality between the matrices thus obtained amounts to identifying the coefficients. One possible implementation is to define a formal matrix consisting of three types of coefficients: 0, 1, *variable*. We defined the following formal computation rules:

$$0 \times 0 = 0$$
$$0 \times 1 = 0$$
$$0 \times variable = 0$$
$$1 \times 0 = 0$$
$$1 \times 1 = 1$$
$$1 \times variable = variable$$
$$variable \times 0 = 0$$
$$variable \times 1 = variable$$

$$0 + 0 = 0$$
$$0 + 1 = 1$$
$$1 + 0 = 1$$
$$0 + variable = variable$$
$$variable + 0 = variable$$

The situations *variable* + *variable*, *variable* + 1, 1 + *variable*, 1 + 1 and *variable* × *variable* never appear, hence the simplicity of the implementation.

The equalities between matrix coefficients will introduce the constraints of matrix structures, with interdependencies. When solving an equation, three scenarios appear: a coefficient must take the value 0, a coefficient must take the value 1 and a coefficient must take the value of another coefficient. An over-constrained system will have no solution.

Finally, we need to resolve the adjacency relations between incidence operators. These can only appear when we have at least two levels of incidence (see surface example). For the adjacency rules between incidence operators, two strategies can be followed. The first, which does not require any implementation work, is to ask users to specify all the impact operators themselves and to ensure their consistency. The advantage for users is that they will not have to specify the adjacency relations between incidence operators. The downside is that this consistency can be tedious to establish.

The second option is to automatically infer incidence operators, depending on the dimensions of the barycentric spaces and adjacency relations between incidence operators. We present the general principle only, the full details of the algorithm can be found in Gouaty (2009). This construction can be done automatically based on the dimensions of the spaces and the adjacency relations between incidence operators. To be able to do this construction automatically, we must determine for each state, the dimension of the associated barycentric space, and especially for each dimension, what the associated control point is. It is then convenient to specify the number of *internal dimensions* for each cell, that is to say, the number of dimensions corresponding to control points, which will influence the geometry of the cell without influencing those of its edges. In the example in section 1.3.2, the curves incident to the face have an internal dimension equal to 1. The control point of the axis e_1 does not influence the position of the vertices. The face has an internal dimension of zero because each control point influences the geometry of at least one edge.

The principle of the algorithm is to build the associated barycentric space (its dimension) and the incidence operators for each state representing a cell. This algorithm needs to be applied recursively:

– the dimension of the barycentric space is initialized to the internal dimension of the cell;

– we add thereto the dimensions of the spaces associated with the subcells;

– the dimension of each subcell should be removed as many times as their area adjacency relations on incidence operators.

Therefore, for each cell, we use the dimension of the subcells. This algorithm has to be applied in a recursive way. The first computed dimension will be that of the "lower level" cells, that is those that do not have an incident cell, in other words the vertices.

Finally, from this first phase and considering the incidence and adjacency relations on incidence operators, it is possible to determine the matrix representations of the incidence operators.

REMARK.– In this section, we have defined cells as attractors of IFS, for the sake of simplicity. However, they can be defined without restriction to C-IFS attractors. In the automaton in Figure 2.21, the edges (the attractors of states a and b) mutually refer to one another. These edges are therefore defined by the sub-automaton consisting of the two states a and b.

1.3.10. *Control of the local topology*

The geometry produced by a BC-IFS depends solely on subdivision operators. The control is not completely free because self-similarity implies that modifying a transformation impacts geometry at every scale level. However, equivalence relations guarantee the preservation of the topological structure induced by incidence and adjacency relations. Regardless of the values of the (free) coefficients of subdivision matrices, the topological structure remains the same. The BC-IFS model constructs a morphism of the parameter space (defined by the symbolic dynamic system enhanced by the equivalence relations on the finite words) toward the attractor. However, the projection following the control points may generate self-intersections, turning the projected attractor non-homeomorphic to the symbolic attractor. This problem appears for many representation models, as is the case with NURBS and subdivision surfaces.

1.3.11. *Principle*

The control of this geometry is done by the choice of coefficients of subdivision matrices (while taking into account the constraints induced by incidence and adjacency equations). As we mentioned in section 1.3.1.2, these matrices can have two interpretations: either as subdivision matrices of the barycentric space, or as refinements of the structure of control points (as for subdivision surfaces). Thus, by achieving the left-side product of the subdivision matrix T by the matrix of the m control points, we obtain m new points, which we call *subdivision points*, ($Q = PT$). Figure 1.57 presents the same surface area as Figure 1.54, with the network of control points in red and the network of subdivision points in blue (obtained by sub-application of subdivision matrices at control points). To build the subdivision matrixes, the barycentric coordinates of the subdivision points have to simply be expressed according to the control points. It is possible to do this from a graphical interface by positioning or moving the subdivision points in the 2D or 3D modeling space. Based on the new positions of the subdivision points, expressed as barycenters of control points, we deduce subdivision matrixes. However, this barycentric expression is not unique, since generally the number of control points (of $\mathbb{R}^3$ is greater than four). To find a unique solution, it is then necessary to use an additional criterion to optimize. To avoid the discontinuity of solutions when we move subdivision points interactively, we have chosen to minimize the distance with the old solution: $||Q - Q_0||$, for example, $Q = Q_0 + T^{-1}(P - P_0)$ where T^{-1} represents the Moore–Penrose pseudo-inverse. The downside of this approach is that we do not have control over each coefficient of each weighted sum. It is sometimes useful to be able to specify them individually.

To fully understand the role and the usage of subdivision points, let us take the example of the curve defined in section 1.3.8. Remember that this curve is defined from three control points, two subdivision operators T_0^C and T_1^C and is bordered by two vertices, each depending on a single control point. Let us choose $\mathbb{R}^2$ as a modeling space. We denote by $P = \begin{pmatrix} P_0|_x & P_1|_x & P_2|_x \\ P_0|_y & P_1|_y & P_2|_y \end{pmatrix}$ the projection matrix, consisting of the three control points P_0, P_1 and P_2. The first three subdivision points are obtained by the matrix product PT_0^C and the other three by $PT_1^C 1$. The incidence and adjacency relations on this curve configuration impose the following subdivision matrix structure:

$$T_0^C = \begin{pmatrix} 1 & \bullet & x \\ 0 & \bullet & y \\ 0 & \bullet & z \end{pmatrix} ; \ T_1^C = \begin{pmatrix} x & \bullet & 0 \\ y & \bullet & 0 \\ z & \bullet & 1 \end{pmatrix}$$

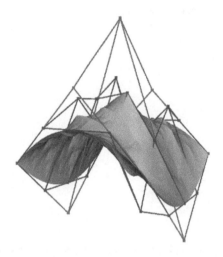

Figure 1.57. *Example of the quadrangular surface bordered by Bezier curves with the network of control points in red and the network of subdivision points in blue. For a color version of this figure, see www.iste.co.uk/gentil/geometric.zip*

As a result, we find that:

– the first subdivision point is identical to P_0;

– the second is arbitrary;

– the third must be the same as the fourth, since the last column of T_0^C is identical to the first column of T_1^C;

– the fifth is arbitrary;

– the sixth is identical to the third control point.

This leaves three subdivision points to be specified to control the local geometry of this curve. Figure 1.58 shows the curves obtained for identical control points and different subdivision points. It should be noted that in this case, the barycentric coordinates of the subdivision points are determined without ambivalence, since they are expressed according to three control points in the plane. Another set of examples is given by Figure 1.59. These curves are defined from three subdivision operators allowing greater freedom in the expression of geometries. The overall geometry is always controlled from the same three control points as in the previous example. Finally, in the last set of examples (see Figure 1.60), we have chosen to further increase the possibilities of form expression, imposing two internal dimensions for curve-type and vertex-type cells. This is represented by 6×6 subdivision matrices, but with matrix overlays by sets of two columns.

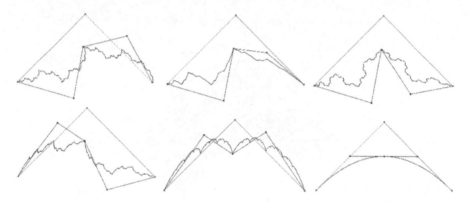

Figure 1.58. *Example of curves projected into the modeling space $\mathbb{R}^2$, following the same control points (in red). The different curves are obtained with different subdivision points (in blue). For a color version of this figure, see www.iste.co.uk/ gentil/geometric.zip*

Figure 1.59. *Example of curves projected into the modeling space $\mathbb{R}^2$, following the same three control points (in red). Each curve is defined by three subdivision operators. The different curves are obtained with different subdivision points (in blue). For a color version of this figure, see www.iste.co.uk/gentil/geometric.zip*

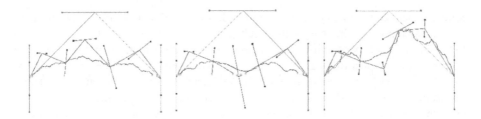

Figure 1.60. *Example of curves projected into the modeling space* $\mathbb{R}^2$, *following the same control points (in red). Each curve is defined by three subdivision operators. Curve-type and vertex-type cells are defined with two internal dimensions. The different geometries are obtained with different subdivision points (in blue). For a color version of this figure, see www.iste.co.uk/gentil/geometric.zip*

1.3.12. *Implementation*

Concerning the implementation, there is nothing specific, since the coefficients of the subdivision matrices merely have to be modified, to change the geometry. However, it is possible to build interactive user interfaces in order to suggest a more intuitive modification of subdivision operators using subdivision points. For this purpose, the control and subdivision points simply have to be visualized, and one has to provide an interface to move them interactively and express these displacements in the subdivision operators, as explained above.

Figure 1.57 presents a network of control points structured according to a mesh, as for subdivision surfaces. Nevertheless, at no time in the BC-IFS model do we explain this structuring. In effect, the control points are only elements allowing us to make the projection of the attractor (defined within the barycentric space) towards the modeling space. We only need to specify which control point corresponds to which dimension. This mapping can be done in two ways. The first solution implicitly defines this mapping. The control point which ought to map the ith barycentric coordinate will constitute the ith column of the projection matrix P. The second solution explicitly formalizes this mapping by an injection between the set of the dimensions of the barycentric space and the set of control points: $\forall j \in J, \exists! i \ tq \ \ I(j) = P_i$. To facilitate the implementation, this injection can be defined for the barycentric space of each state q. In both cases, this mapping must be defined for each state directly accessible from the initial state corresponding to the attractor projected into the modeling space. For the other states, the mapping will be automatically inferred from this initial definition and from the embedding operators.

However, once this mapping is established, based on incidence and adjacency relations on incidence operators, we can deduce a structure between the control points. Incidence relations define a hierarchy between cells. This hierarchy results in an interweaving of the barycentric subspaces associated with the cells. Each dimension of each barycentric space is associated with a control point, so we can infer a hierarchy or a tree-based structure on the control points. However, we must also take the adjacency constraints on the incidence operators into account. These constraints induce the sharing of dimensions between the barycentric spaces of the cells under consideration. For example, two adjacent edges of the same face have a common vertex. Thereby, the spaces of these two edges will have a dimension corresponding to the same vertex. Then, the tree-based structure, determined from incidence relations, will develop into a graph due to the adjacency constraints on the incidence operators.

Different alternatives can be used to visualize this structure. One of the simplest is to construct the graph representing the relationships between the cells as follows:

– for each cell, we determine the barycenter of the control points associated with the internal dimensions;

– each of these control points is connected to the barycenters;

– for each incidence relation, the subcell barycenter is connected to that of the main cell.

Figures 1.57–1.61 show this structure. For the surface shown in Figure 1.57 and the curves in Figures 1.58 and 1.59, each cell has an internal dimension equal to 1. The barycenters are then merged with the corresponding control points. The curve in Figure 1.60 brings forward the internal control points (here each cell has an internal dimension equal to 2). The choice to not order internal control points is deliberate as they have no relation, and they influence the geometry of the cell independently. The structure of subdivision points is constructed through the image of the control point structure by application of subdivision operators. We thus obtain as many structures as operators. These structures are interconnected according to the adjacency relations. Figure 1.61 shows the case of a triangular face, with an internal dimension equal to 3 for each cell (face, edges and vertices).

1.3.13. *Choice of the display primitive*

As we saw during the presentation of the IFS, the choice of the display primitive has no influence on the attractor, as it is the fixed point to which any sequence built by iterating the Hutchinson operator converges from any initial primitive. However, in practice, the evaluation of the attractor is always made from a finite number of iterations. The choice of the display primitive then becomes important, or even essential, when it comes to calculating an approximation of the attractor in order to materialize it using 3D printing. In this case, the object approximating the attractor must be consistent (with a non-empty and connected interior), even if the attractor itself does not have an interior. For example, the Menger sponge is a 3D object of topological dimension 1 with an empty interior. In absolute terms, it cannot be materialized. Nonetheless, we can easily build a series of solid objects approximating the Menger sponge, choosing the cube encompassing the attractor as the primitive.

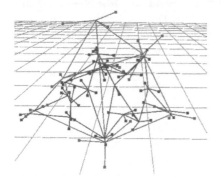 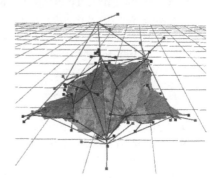

Figure 1.61. *Example of a network of control points for a triangular surface that subdivides into four parts. The internal dimension of each cell (face, edges and vertices) is equal to 3. The left-hand side illustration only shows the network of control points and subdivision points, the resulting surface has been added to the right-side illustration. For a color version of this figure, see www.iste.co.uk/gentil/geometric.zip*

Whether we are trying to approximate the attractor of an IFS, a C-IFS or a BC-IFS, the choice of the primitive can be guided by the criteria outlined below.

1.3.14. *Choice of dimension*

It seems relevant and intuitive to choose an object with the same dimension as the attractor as the primitive: for an attractor such as a curve or a wired structure, we will choose a line or curve, for a surface, a facet or a set of facets, etc. This choice is not always relevant when the fractal dimension is greater than the topological dimension. For the Sierpinski triangle (respectively, the Menger sponge), of topological dimension 1 (respectively, 3) a triangle (respectively, a cube) will be better suited. Consideration must also be given to the expected exploitation of the approximation. To make the approximation of the attractor, a volume will be necessary. A primitive volume will be required. We will have to perform post-processing with the approximation, such as the computation of a 3D offset or an extrusion, etc.

1.3.15. *Choice of geometry*

The closer the initial primitive is to the attractor, the faster the sequence of iterations by the Hutchinson operator will converge. For example, for the Sierpinski triangle, the triangle whose vertices correspond to those of the Sierpinsky triangle will constitute a particularly relevant primitive.

A very good compromise to automatically determine an initial primitive is to build it from the set of fixed points of the IFS transformations (broken line connecting the fixed points, convex envelope of the fixed points, etc.). These fixed points are a part of the attractor, but are also a part of all their iterations. Therefore, the figure built by iteration is a figure whose vertices are a part of the attractor.

1.3.16. *Primitive property*

The primitive can verify a certain number of properties that can subsequently be employed for the calculation of an approximation of the attractor or a geometric operation with the attractor (such as its intersection with a straight line for ray tracing visualization). These properties are presented here in the case of IFS, but can simply be generalized to C-IFS and BC-IFS.

DEFINITION.– *A primitive K verifies the property of rapid decay with respect to an IFS $\mathbb{T}$ if:*

$$\mathbb{T}(K) \subset K$$

PROPERTY.– If the primitive K verifies the decay property in relation to an IFS $\mathbb{T}$, then the sequence $\mathbb{T}^n(K)$ is decreasing and $\forall n \in \mathbb{N}, \mathcal{A}(\mathbb{T}) \subset \mathbb{T}^n(K)$.

The sequence $\mathbb{T}^n(K)$ then constitutes a decreasing sequence encompassing the attractor. The other interest of this approach is to have an approximation that possesses, at most, the same number of connected components as the attractor. If two parts of the attractor intersect $(T_i(\mathcal{A}) \cup T_j(\mathcal{A}) \neq \emptyset)$, the same holds for $\mathbb{T}(K), \forall K \supseteq \mathcal{A}$, that is to say, $T_i(K) \cup T_j(K) \neq \emptyset$. This type of primitive can be adapted to visualize the attractor, but it can cause problems if we need to give a geometric representation of the result. Non-empty intersections will induce more or less complex treatments to determine the result of the union of the two parties involved. Imagine a primitive represented by a mesh. The calculation of the union of the transformations of this mesh, by the transformations of the IFS, can become computationally expensive to evaluate after a few iterations.

1.3.17. *BC-IFS*

Here, we discuss the specific case of BC-IFS, whose attractors are defined in a barycentric space. If an attractor is made up of several parts involving multiple states, we must define one primitive per state, described in the associated barycentric space.

The second element to consider is that the topology of BC-IFS is determined by incidence and adjacency relations.

As mentioned earlier for IFS, it is relevant to build a display primitive from the fixed points of the transformations. For transformations defined in the barycentric space, these fixed points correspond to the eigenvectors associated with the eigenvalue 1. From this set of fixed points, we can automatically build the topological structure of the primitive by employing incidence relations and adjacency relations on incidence operators. Indeed, these relations reflect the topological structure of the different levels of cells and of their interweaving. One of the interests of this construction is that the connections between the images of the primitives through the transformations of the BC-IFS will be in accordance with the adjacency relations. In fact, the connections induced by adjacency relations are defined between subcells up to the cells of the lowest level which are vertices, that is, the fixed points of transformations. Therefore, primitives built from fixed points will properly connect during the iteration process. However, automatic construction also has its limits and it is sometimes necessary to define these primitives manually.

2

Design Examples

The formalism of BC-IFS offers remarkable modeling possibilities. In this chapter, we present a number of pedagogical examples, highlighting these possibilities and showing how it is possible to "play" with the different concepts related to BC-IFS to design and control forms of different kinds.

2.1. Curves

Curves are an excellent introduction for understanding the influence of the parameters of BC-IFS and how we can make use of them. First, curves do not have any fractal topological structure. They have a dimension of 1. Only the geometric structure can be fractal.

To build a curve, we have to choose how many transformations we want to use to describe it. This choice determines the number of self-similar parts that will make up the curve. The more transformations we use, the more degrees of freedom we shall have to change its geometry.

We shall start with curves whose automaton is represented by Figure 2.1. We shall see at the end of this section that other automatons are possible. In the meantime, we need to specify the number of subdivision transformations, as well as the number of control points that we want to assign to each topological cell. For a curve, cells are the curve itself and the two vertices that border it. The number of control points for the curve defines the number of degrees of freedom that we shall have to change the overall geometry of the curve. It also induces the number of degrees of freedom available for defining the geometry of each self-similar part (subdivision points = image of control points by every transformation). Figure 2.2 shows two curves defined from two transformations. The first is defined by four control points, and the second is defined by five control points.

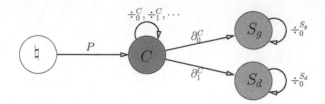

Figure 2.1. *"Standard" automaton for a curve. Transitions referred to by the symbol ÷ represent subdivision operators, while those referred to by the symbol ∂ represent incidence operators. The symbol P represents the projection operator of the attractor (defined within a barycentric space) into the modeling space. For a color version of this figure, see www.iste.co.uk/gentil/geometric.zip*

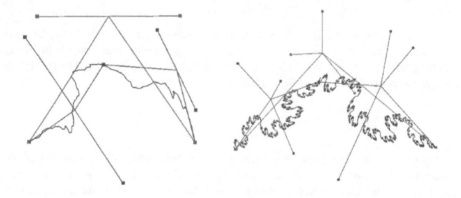

Figure 2.2. *Examples of curves constructed using two transformations. The control points are shown in red and the subdivision points are in blue. For both curves, each vertex depends on a single control point: on the left side the curve depends on four control points, and on the right side the curve depends on five control points. For a color version of this figure, see www.iste.co.uk/gentil/geometric.zip*

In these examples, incidence and adjacency constraints are those presented in section 1.3, that is:

$$\partial_0^C \div_0^{S_g} \simeq \div_0^C \partial_0^C$$
$$\partial_1^C \div_0^{S_d} \simeq \div_1^C \partial_1^C$$
$$\div_0^C \partial_1^C \simeq \div_1^C \partial_0^C$$

It is possible to define another type of connection. The tree in Figure 2.3 shows an example, where the differences with a "standard"-type connection are in red. We can see that the right-hand subdivision $\div_1^C$ is "reversed" (compared to the original curve), and it is the right-side edge ∂_1^C of this subdivision that is connected to the right-side edge ∂_1^C of the other subdivision $\div_0^C$. The vertices of the right- and left-hand sides must always be of the same kind (with the same attractor, that is, of the same state) but this time this constraint is imposed by the incidence equation (equation [2.2] hereafter) and not by the adjacency equation (as in the "standard" case). Incidence and adjacency constraints are of the form:

$$\partial_0^C \div_0^{S_g} \simeq \div_0^C \partial_0^C \qquad [2.1]$$

$$\partial_1^C \div_0^{S_d} \simeq \div_1^C \partial_0^C \qquad [2.2]$$

$$\div_0^C \partial_1^C \simeq \div_1^C \partial_1^C \qquad [2.3]$$

The family of curves generated by this type of connection is different from that generated by the standard connection. Figure 2.4 shows three examples that are compared to the curves constructed from two BC-IFS, $T^A = \{T_0^A, T_1^A\}$ and $T^B = \{T_0^B, T_1^B\}$. The connections defined for T^A are standard and those of T^B are reversed. The same subdivision points are used in both cases. Even if the subdivision points are the same, T_1^A and T_1^B are different. The first and third columns of T_1^B are permutated with respect to those of T_1^A due to the inversion of the second self-similar portion connected to T_1^B. On the other hand, T_0^A and T_0^B are identical. In Figure 2.4, we can see three examples for each of the two types of connection. For the inverted connection, we can see that the second part of the curve is copied by "turning it upside down". The third column shows, at the bottom, the example of the dragon curve, which includes one inversion; at the top, the corresponding curve without inversion.

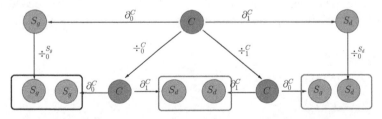

Figure 2.3. *A tree illustrating another possibility for connecting the subdivisions of a curve. The right-side subdivision is inverted. For a color version of this figure, see www.iste.co.uk/gentil/geometric.zip*

These constructions can be enhanced almost indefinitely by increasing the number of transformations. The two types of connection can then be combined at will for each subdivision.

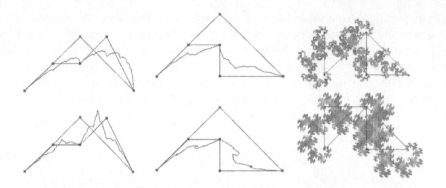

Figure 2.4. *Comparison of the effect of the two types of connection, standard and inverted, on curves constructed from two transformations. For a color version of this figure, see www.iste.co.uk/gentil/geometric.zip*

COMMENTS ON FIGURE 2.4.– *The vertices depend on one control point and the curves on three. The images on the top line represent curves constructed using standard connections; those of the bottom line, curves with an inverted connection. The image at the bottom right shows an example of the dragon curve, built from an inverted connection. The equivalent curve, without an inverted connection, is shown top right.*

Figure 2.5 presents four examples of curves defined from three subdivisions. Figure 2.6 illustrates how the number of transformations influences the control of the curve. For this example, we choose to build two curves, both depending on three control points. We have defined the von Koch curve using two BC-IFS: the first is composed of two transformations of $\mathbb{T}^A = \{T_0^A, T_1^A\}$, and the second of four $\mathbb{T}^B = \{T_0^B, T_1^B, T_2^B, T_3^B\}$. We chose the transformations in such a way that:

$$T_0^B = T_0^A T_0^A$$
$$T_1^B = T_0^A T_1^A$$
$$T_2^B = T_1^B T_0^A$$
$$T_3^B = T_1^A T_1^A$$

We denote this by $\mathbb{T}^B = (\mathbb{T}^A)^2$. Therefore, the two BC-IFSs have the same attractor, as can be seen in the left column of Figure 2.6. For these two examples, incidence and adjacency constraints have been defined in a standard way (that is, without inversion). Displacing control points identically affects the curves because:

– on the one hand, the attractors are identical (both attractors are defined in barycentric spaces of the same dimension, here equal to 3);

– on the other hand, control points define the projection of the attractor in the modeling space.

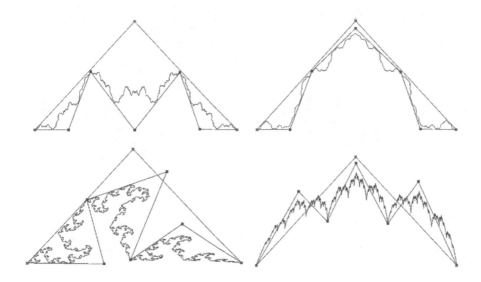

Figure 2.5. *Examples of curves constructed using three transformations whose vertices depend on a single control point, and the curve depends on three control points. For a color version of this figure, see www.iste.co.uk/gentil/geometric.zip*

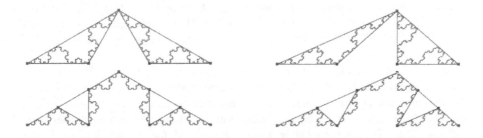

Figure 2.6. *Examples of von Koch curves constructed with two transformations (top line) and four transformations (bottom line). For the left column, the control points have been chosen to obtain a projection that corresponds exactly to the von Koch curve. The right-hand column illustrates the effect of moving the second control point, thus changing the projection of the attractor. For a color version of this figure, see www.iste.co.uk/gentil/geometric.zip*

Figure 2.7 highlights the differences in controlling the geometry between $\mathbb{T}^A$ and $\mathbb{T}^B$.

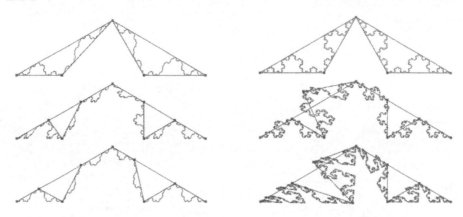

Figure   2.7. *Left-hand   column: for   the   first   figure,   the   first subdivision   point   of $\mathbb{T}^A$ has   been   moved,   thus   changing   the geometry of the attractor. For a color version of this figure, see www.iste.co.uk/gentil/ geometric.zip*

COMMENTS ON FIGURE 2.7.– *For the second figure, the second subdivision point of $\mathbb{T}^B$ (similar to the first of $\mathbb{T}^A$ has undergone the same displacement. The geometry of the attractor is modified, though not in the same way as the figure above. The third figure shows that it is possible to obtain the attractor of $\mathbb{T}^A$ (first line); to this end, one merely has to choose the subdivision points such that $\mathbb{T}^B = (\mathbb{T}^A)^2$. Right-hand column: at the top, as a reference, the von Koch curve built using $\mathbb{T}^A$. Presented underneath are different curves obtained using $\mathbb{T}^B$. These curves cannot be generated using $\mathbb{T}^A$.*

So far, we have built curves involving a single curve-type state. It is possible, however, to define curves that subdivide into several curves of different kinds, each with its own characteristics (number of control points, number of transformations). For example, a curve CA can be subdivided into a CA-type curve and another into a CB-type. The CB-type curve can in turn be subdivided, for example into a CB-type curve and a CA-type curve. The automaton on the left side of Figure 2.8 represents this type of subdivision. On the right, we show the result of the execution of the automaton for each curve type. Incidence constraints are shown in blue and adjacency constraints in red.

These constraints are expressed by:

– for the CA-type curve:

$$\partial_0^{CA} \overset{\cdot}{\div}_0^{SA_g} \simeq \overset{\cdot}{\div}_0^{CA} \partial_0^{CA}$$

$$\partial_1^{CA} \overset{\cdot}{\div}_0^{SA_d} \simeq \overset{\cdot}{\div}_1^{CA} \partial_1^{CB}$$

$$\overset{\cdot}{\div}_0^{CA} \partial_1^{CA} \simeq \overset{\cdot}{\div}_1^{CA} \partial_0^{CB}$$

– for the CB-type curve:

$$\partial_0^{CB} \overset{\cdot}{\div}_0^{SB_g} \simeq \overset{\cdot}{\div}_0^{CB} \partial_0^{CB}$$

$$\partial_1^{CB} \overset{\cdot}{\div}_0^{SB_d} \simeq \overset{\cdot}{\div}_1^{CB} \partial_1^{CA}$$

$$\overset{\cdot}{\div}_0^{CB} \partial_1^{CB} \simeq \overset{\cdot}{\div}_1^{CB} \partial_0^{CA}$$

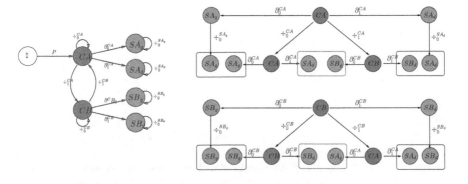

Figure 2.8. *Automaton describing a CA curve subdivided into a CA-type curve and another CB-type curve. This second curve is subdivided into a CA-type curve and a CB-type curve. For a color version of this figure, see www.iste.co.uk/gentil/ geometric.zip*

An example of such a curve is given in Figure 2.9.

2.2. Wired structures

The connections between the subdivisions of a curve are defined so as to form a chain going from one vertex to another, with or without inversions, and eventually involving different types of curves. It is possible, however, to define other types of connections between subdivisions. In particular, by achieving "loops", as shown by the tree in Figure 2.10. Hereafter, the expressions of corresponding incidence and adjacency relations are as follows:

$$\partial_0^C \div_0^{S_g} \simeq \div_0^C \partial_0^C \tag{2.4}$$

$$\partial_1^C \div_0^{S_d} \simeq \div_1^C \partial_1^C \tag{2.5}$$

$$\div_0^C \partial_1^C \simeq \div_1^C \partial_0^C \tag{2.6}$$

$$\partial_0^C \div_0^{S_g} \simeq \div_2^C \partial_0^C \tag{2.7}$$

$$\div_0^C \partial_1^C \simeq \div_2^C \partial_0^C \tag{2.8}$$

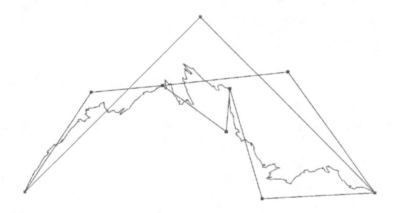

Figure 2.9. *Example of a curve obtained with two mutually referenced states according to the automaton described in Figure 2.8. For a color version of this figure, see www.iste.co.uk/gentil/geometric.zip*

In this set of relations, we find a first set of constraints with two incidence relations [2.4] and [2.5] and an adjacency relation [2.6] defining a curve. This curve is the sub-attractor defined by the subdivisions $\div_0^C$ and $\div_1^C$. We can also identify a second set of relations (two incidence relations [2.4] and [2.7], as well as an adjacency relation [2.8] defining a second curve (sub-attractor) from the subdivisions $\div_0^C$ and $\div_2^C$. Figure 2.11 shows two examples of such structures. The two curves identified by the incidence and adjacency relations are represented in the second and third lines. The attractors of the first line are not made up of the simple union of the two curves, all the interactions between the transformations must be added. We can see the fractal topological structure described by our constraints, on the left with a "loop" consisting of two copies of the shape and, on the right, the same duplicated and concatenated form. The structure is obviously repeated indefinitely according to the principle of self-similarity.

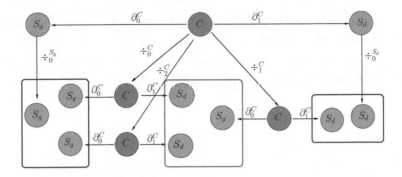

Figure 2.10. *Tree illustrating the construction of a connection for a wired structure. For a color version of this figure, see www.iste.co.uk/gentil/geometric.zip*

Other examples of the construction of wired structures are presented in Figure 2.12.

2.3. Surfaces and laces

The construction of surfaces is similar to that of curves, with an additional incidence level between cells. It brings forward the structuring of objects into cells and subcells and the distribution of control points in the cell hierarchy.

This construction must comply with a certain number of rules:

– the surface should be bounded by curves. The latter must therefore satisfy the incidence and adjacency constraints of a curve, otherwise, the face would have a fractal topology bounded by Cantor sets;

– the surface subdivisions must be connected following the edges (adjacency relation) so as to constitute a "solid" and consistent surface, in other words, one without a hole or a dangling face;

– the orientations of the curves must be respected to avoid incompatible constraints. This point will be developed further in the text (see section 4.1);

– subdivision operators must be contractive.

Therefore, the design of self-similar surfaces is naturally connected to the issue of the tessellation of the plane. We then find the standard constructions: triangular, quadrangular, hexagonal subdivision, etc. However, we have a few additional constraints:

– the surface to be tiled is bounded;

– the structure of this surface is imposed or chosen, but fixed (that is, the number of edges, the nature of the edges);

– the edges at which the elements are connected must be compatible (same number of subdivisions, same barycentric space and same orientation).

Figure 2.11. *Example of two wired structures built from the incidence and adjacency relations defined by equations [2.4]–[2.8]. For a color version of this figure, see www.iste.co.uk/gentil/geometric.zip*

We shall illustrate these different aspects with the following examples.

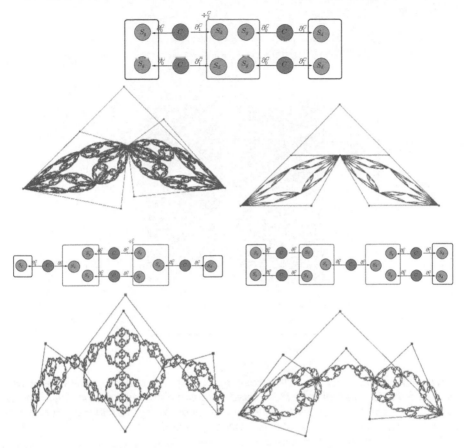

Figure 2.12. *Examples of wired structures. The diagrams above each shape represent the first level of subdivision of the tree symbolizing the execution of the BC-IFS automaton. For a color version of this figure, see www.iste.co.uk/ gentil/geometric.zip*

COMMENTS ON FIGURE 2.12.– *Incidence relations are indicated in blue and adjacency relations are in red. At the top, the BC-IFS is built from four transformations. The subdivided cells are linked two by two to form two successive loops. On the left, the attractor represents a first instantiation of transformations. On the right, the transformations have been chosen to obtain Bezier curves. At the bottom, other structures with fractal topology are represented. On the left: a loop built from two subdivisions is surrounded by two subdivisions. On the right: this time, a subdivision in the center is surrounded by two loops, each built from two subdivisions.*

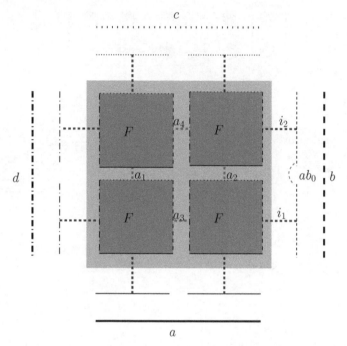

Figure 2.13. *Diagram of the cellular decomposition of a quadrangular surface with incidence relations indicated by blue dotted lines and adjacency relations indicated by red dotted lines (a_1, a_2, a_3 and a_4 for the face and ab_0 for the curve b). For a color version of this figure, see www.iste.co.uk/gentil/geometric.zip*

2.3.1. *Quadrangular subdivisions*

First, let us take the example of the construction of a proposed quadrangular surface, as in section 1.3 and described by Figure 1.52. We simplify this representation by eliminating the symbols and by representing the edges with different lines to obtain Figure 2.13. We note that if we do not make any assumptions about the constraints imposed on the edges, we can define four different types of edges a, b, c and d (that is, the four different states). However, the adjacency constraints identified by a_1 and a_2 in Figure 2.13 require the states a and c to be the same. Similarly, adjacency constraints a_3 and a_4 require that states b and d be the same. So for this type of configuration, we can have at most two different types of edges: a type corresponding to the edge connecting at a_1 and a_2 and a type of curve connecting at a_3 and a_4.

The set of these constraints explained in section 1.3.8 is sufficient but not necessary. There is redundancy in these relations. For example, it is not necessary to specify adjacency relations for each curve, since the incidence relations of curves and adjacency relations between faces induce the adjacency relations of curves. We can easily understand this from Figure 2.13. the adjacency ab_0 is inferred from the two incidence relations i_1 and i_2 and the adjacency relation a_2. Similarly, it is possible to deduce from other relations. For example, it is possible to deduce equivalency relations highlighting two curve structures that run diagonally across the surface, by considering the two subsystems (sub-BC-IFS) restricted to two subdivisions arranged diagonally.

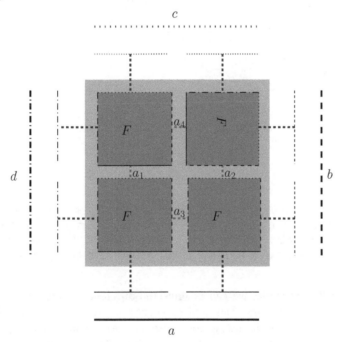

Figure 2.14. *Example of subdivision of a quadrangular surface with "non-standard" connections. The top right face has undergone a rotation of 90°. Incidence and adjacency relations are always written in the same way but do not involve the same symbols. For a color version of this figure, see www.iste.co.uk/gentil/geometric.zip*

For curves, we had defined two types of connections referred to as standard and inverted. Similarly, for surfaces, we can make other choices of connections between the subdivided faces. Figure 2.14 shows another configuration where the top right face has been rotated by 90°. With this configuration, the topology is still that of a quadrangular surface, the geometry will be copied in a different manner during the iterations, and the final attractor will be different. The connection conditions are no longer defined on the same pair of edges as before. The a_1 constraint imposes $a = c$ as before, but a_2 imposes $c = b$. The constraint a_3 results in $d = c$ and a_4 results in $b = a$. In the end, this configuration requires only one type of edge.

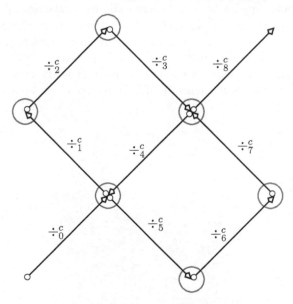

Figure 2.15. *Connections to build a Hilbert/Peano curve. The red circles show the adjacency relations that we can define to bring us closer to the definition of a Hilbert/Peano curve. However, these relations are insufficient to ensure that the curve fills a surface. For a color version of this figure, see www.iste.co.uk/gentil/geometric.zip*

Following this principle, we can imagine the different configurations, we could build using rotations or symmetries applied to either subdivision. We can also adjust the number of subdivisions following both directions. An example of a combination of these two parameters is the construction of the Hilbert/Peano "curve". It can be built by treating it as a curve that subdivides into 9 parts. However, in this case it will not be possible to modify the geometry of the curve based on the subdivision points while ensuring that a surface is filled. The principle of building this type of curve implies additional connections. Figure 2.15 shows the adjacency relations that can be defined from a curve that is subdivided into 9 parts. Nonetheless, these connections are not sufficient to guarantee that the final curve fills the space because we do not define the topology of a surface. It is possible to fill a surface but by accurately choosing subdivision operators and control points, as shown in Figure 2.16.

Figure 2.16. *Example of a curve attempt that satisfies the adjacency relations in Figure 2.15. For a color version of this figure, see www.iste.co.uk/gentil/geometric.zip*

COMMENTS ON FIGURE 2.16.– *The first row shows iterations 1, 2 and 4 of a curve that satisfies adjacency relations, but for which subdivision points are not correctly adjusted to achieve a curve that fills a surface. The image of the second line on the left shows an adjustment of the subdivision points to obtain the Hilbert/Peano curve (other symmetrical configurations produce a similar result). On the right: the slightest deviation in the position of one of the subdivision points produces "tearing" or surface overlays.*

If we design the Hilbert/Peano curve based on the cellular decomposition of a curve, we cannot define all of the constraints necessary to guarantee the construction of a surface. If we utilize the subdivision of a surface then all constraints can be explained, as shown in Figure 2.17.

Visualization from a primitive of the quadrangular face type does not reveal the structure of the Hilbert/Peano curve because we obtain a surface at each iteration. Nevertheless, by closely observing the geometry, it is possible to perceive this structure (see Figure 2.18).

The BC-IFS formulation is very general and allows for the description of many iterative processes. Here, we give an example showing the construction of a quadrangular surface from only two subdivisions. Figure 2.20 shows how it is possible to achieve this construction. Implementing this two-dimensional solution is very easy given the connection problems are trivial, especially when the edges are line segments (as shown in Figure 2.20). When the edges are curves, it becomes more complex. The use of Bezier curves can be relatively simple to implement because they are symmetrical and do not require the issue of orientation to be addressed.

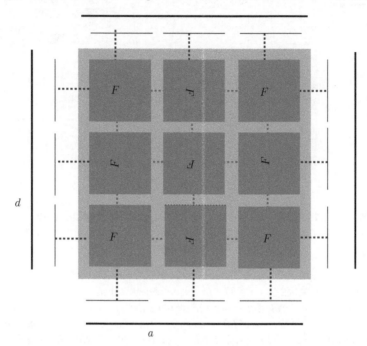

Figure 2.17. *Subdivision of a quadrangular surface satisfying the constraints of a Hilbert/Peano surface. Each subdivided face is oriented so as to follow the path of the Hilbert/Peano curve. For a color version of this figure, see www.iste.co.uk/ gentil/geometric.zip*

However, if we want to achieve connections to generate a consistent geometric pattern, or simply make connections following self-similar curves that are not necessarily symmetrical, "orientation constraints" must be taken into account. Therefore, the concept of orientation constraints is important in the definition of connections. So far, we have dealt with it but not explained it. This will be developed in section 4.1 by proposing solutions to facilitate the design phase. For now, we shall intuitively address it with an example. In the context of BC-IFS, when we talk about orientation, it is a question of identifying where the edges of a cell are located, relative to one another.

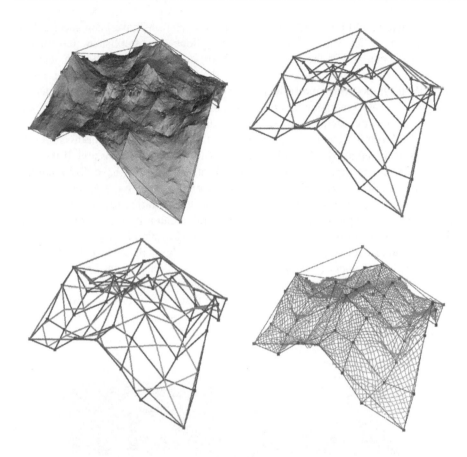

Figure 2.18. *Example of a Hilbert/Peano surface, defined from the subdivision of a surface with the incidence and adjacency relations presented in Figure 2.17. For a color version of this figure, see www.iste.co.uk/gentil/geometric.zip*

COMMENTS ON FIGURE 2.18.– *At the top left, the surface is visualized using a quadrangular surface primitive and for five iterations. For each iteration, we obtain a quadrangular surface. For the other illustrations, the surface is visualized using a wired primitive (line segment). Iterations 1, 2 and 4 are presented successively. At the first iteration, we perceive in green the path of the Hilbert/Peano curve.*

Let us begin again from the example of the subdivision of a quadrangular face, presented in Figure 1.52. Let us look into the curve a at the bottom of the figure. We have assigned two vertex-type edges thereto, arbitrarily arranged: on the left side for the edge ∂_0^a and on the right for the edge ∂_1^a. For reasons of consistency in the notation, the subdivisions $\div_0^a$ and $\div_1^a$ have been placed on the left- and right-hand sides, respectively, and the incidence and adjacency relations have been specified in this configuration. Once these choices or conventions are established, the matrix coefficients associated with the subdivision $\div_0^a$ will instead influence the geometry of the left side of the curve and those of $\div_1^a$ the right-hand side. Some of these coefficients are imposed by the incidence constraints involving incidence operators. Thus the result is an orientation induced by the incidence operators ∂_0^a and ∂_1^a. When defining an adjacency relation between two subdivided faces (for example, between $\div_1^F$ and $\div_2^F$) through a curve a, we impose that the edge ∂_0^F of the subdivision $\div_2^F$ corresponds to the edge ∂_0^F of subdivision $\div_1^F$. This will only be possible if the two edges (curves) of the type a (at the bottom, ∂_0^F at the top, ∂_2^F) have "compatible" orientations. In other words, if after subdivision of the faces by $\div_1^F$ and $\div_2^F$, the edges of these subdivided faces end up with the same orientation (defined by ∂_0^a and ∂_1^a). Without this, the problem may be over-constrained and the system of incidence and adjacency constraints will not have any solution. However, it may be possible to make connections with edges of opposite orientation, this will lead to "twisting" of the face if the latter is not otherwise constrained (see section 4.1). In Figure 1.52, the two type-a curves (top and bottom) have their edges 0 (respectively, 1) to the left (respectively, to the right). Similarly, we can note that the two type-b curves, on the right- and left-hand sides, have their edges 0 (respectively, 1) at the bottom (respectively, at the top).

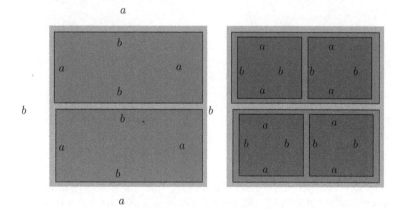

Figure 2.19. *The diagram on the right-hand side presents the quadrangular subdivision in the plane from two subdivisions. The initial square is transformed by a contraction following the x-axis, and then a rotation of 90°. For a color version of this figure, see www.iste.co.uk/gentil/geometric.zip*

COMMENTS ON FIGURE 2.19.– *Two copies are made up of the rectangle thus obtained and the copies are positioned on top of each other. Type-a edges will thus be mapped with one of the halves of the type-b edge. The two transformations thus defined are not contractive but power-contractive. We can observe this from the second iteration on the left diagram where the subdivided elements are squares of sides, twice as small as the original square. This simple IFS, composed of the two transformations, has the original square as the attractor. However, in order to construct a geometric model consistent with more sophisticated edges, it is necessary to define, precisely and coherently, the incidence and adjacency relations (see Figure 2.20).*

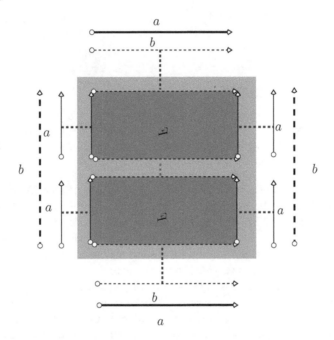

Figure 2.20. *To achieve a quadrangular surface from two subdivisions, we need to explain the connections and find a consistent system for the orientation of the faces and their edges, according to the connections. For a color version of this figure, see www.iste.co.uk/gentil/geometric.zip*

COMMENTS ON FIGURE 2.20.– *To properly achieve the necessary topological connections, it is first necessary to identify the subdivision system of the edges (which is not trivial). The type-a edge is not subdivided during the first iteration, but only at the second. The edge a must be subdivided into a single edge b which is divided into two. Similarly, the edge b must be subdivided in half during the first iteration but not during the next. This is expressed by specifying that b is divided into two edges a. This diagram shows a possible solution. After having defined the edge subdivisions and chosen an orientation, we can connect the edges of the subdivided faces (after rotation) with the edge subdivisions. However, to have the compatibility of the orientations of the edges, we have applied a horizontal symmetry on each subdivided face. The diagram of the automaton describing the subdivision system is shown in Figure 2.21.*

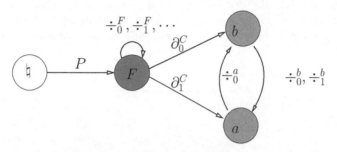

Figure 2.21. *Automaton symbolizing the subdivision system of a quadrangular surface, with two subdivisions. Only the states of the face (F) and the edges (a and b) have been represented. For a color version of this figure, see www.iste.co.uk/gentil/geometric.zip*

We now return to the case of the quadrangular surface defined by two subdivision operators, as shown in Figure 2.19, and define the constraints taking the orientations into account. Figure 2.20 shows the face bounded by four curves, two of type a and two of type b. The sides of the edges and of the subdivisions 0 of these curves are represented by a circle, while the edges and subdivisions 1 are indicated by an arrow. We can see that the b-curve is divided into two a-curves with the "same orientation". The b-curve is divided into a single b-curve of the same orientation. To achieve a coherent subdivision system, the face is subdivided into two sides to which we have applied a rotation of 90°, and a horizontal symmetry, to align the orientations of the edges. These rotations and symmetries are of course symbolic, they just allow us to arrange the elements in relation to one another and to visualize the incidence and adjacency constraints and therefrom facilitate their notation.

A visualization of this type of surface is presented by Figure 2.22, where the curves bounding the face are Bezier curves.

2.3.2. *Triangular subdivisions*

Triangular subdivisions have been extensively studied and employed, particularly for subdivision surfaces, such as the Loop or Butterfly diagrams. The simplest of these subdivisions is presented in Figure 2.23. The translation of this type of topological subdivision using BC-IFS is simple and immediate. Once the cell subdivision is identified, we can write the equivalence relations to define the topology of the whole.

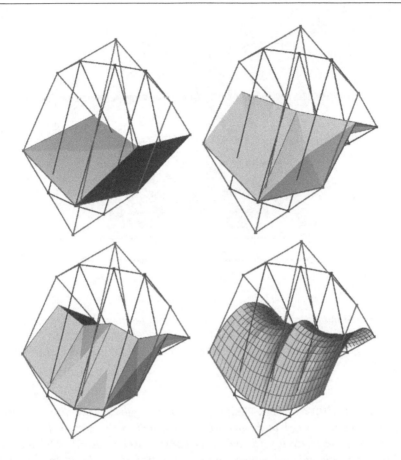

Figure 2.22. *Example of a quadrangular surface with two subdivisions. For this illustration, the curves bounding the surface are Bezier curves. For a color version of this figure, see www.iste.co.uk/gentil/geometric.zip*

The incidence relations are given as:

$$\partial_0^F \div_0^a \simeq \div_0^F \partial_0^F \qquad\qquad [2.9]$$

$$\partial_0^F \div_1^a \simeq \div_1^F \partial_0^F \qquad\qquad [2.10]$$

$$\partial_1^F \div_0^b \simeq \div_1^F \partial_1^F \qquad\qquad [2.11]$$

$$\partial_1^F \div_1^b \simeq \div_2^F \partial_1^F \qquad\qquad [2.12]$$

$$\partial_2^F \div_0^c \simeq \div_2^F \partial_2^F \qquad\qquad [2.13]$$

$$\partial_2^F \div_1^c \simeq \div_0^F \partial_2^F \qquad\qquad [2.14]$$

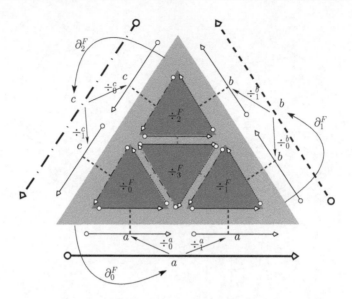

Figure 2.23. *Standard triangular subdivision. The triangular face is subdivided into four triangular sub-faces (indicated by the corresponding symbols of the alphabet $\div_0^F, \div_1^F, \div_2^F$ and $\div_3^F$). For a color version of this figure, see www.iste.co.uk/gentil/geometric.zip*

COMMENTS ON FIGURE 2.23.– *The three "edges" (curves bounding the face) are identified by the letters a, b and c and are represented by the arrows in solid, dotted and mixed lines (respectively). The incidence constraints are identified by links in blue dotted lines and the adjacency constraints by red dotted line links. In this accurate configuration, we find that edges of dotted and mixed types must be of the same kind (that is, defined by the same state $b = c$) due to adjacency constraints. For curves, in order to lighten the diagram, the "right-hand side" vertices (for example, reached by ∂_0^x with $x = a, b$ or c) and "left-hand side" vertices (namely, reached by ∂_1^x with $x = a, b$ or c) are, respectively, represented by an arrow and a circle.*

The adjacency relations are given as:

$$\div_0^F \partial_1^F \simeq \div_3^F \partial_2^F \tag{2.15}$$

$$\div_1^F \partial_2^F \simeq \div_3^F \partial_1^F \tag{2.16}$$

$$\div_2^F \partial_0^F \simeq \div_3^F \partial_0^F \tag{2.17}$$

If we do not go further in the specification of equivalence relations, we are guaranteed to have a "face" bounded by "curves". We should nonetheless be cautious, as the curves in question may be Cantor sets (because we do not have any topological constraints on this type of cell) according to which the subdivisions of the faces are connected.

It is possible to write incidence and adjacency relations for each curve as below, but these are already induced by the previously written equations:

$$\partial_0^x \div_0^S \simeq \div_0^x \partial_0^x \qquad [2.18]$$

$$\partial_1^x \div_0^S \simeq \div_1^x \partial_1^x \qquad [2.19]$$

$$\div_0^x \partial_1^x \simeq \div_1^x \partial_0^F \qquad [2.20]$$

where x successively represents a, b and c.

Finally, we need to specify the adjacency constraints on the incidence relations, expressing that two adjacent curves share a common vertex:

$$\partial_0^F \partial_1^a \simeq \partial_1^F \partial_0^b \qquad [2.21]$$

$$\partial_1^F \partial_1^b \simeq \partial_2^F \partial_0^c \qquad [2.22]$$

$$\partial_2^F \partial_1^a \simeq \partial_0^F \partial_0^a \qquad [2.23]$$

From this first example, even for a standard triangular subdivision, the combinations are multiple. In fact, the side subdivisions have been arranged in such a way to connect the edges of the same kind (for example, a type a edge with a type a edge). However, it is possible to arrange them differently, as illustrated by Figure 2.24.

2.3.3. *Other regular polygonal subdivisions*

Without going into too much detail, we present some subdivisions of regular polygonal surfaces: Figure 2.25 shows a subdivision of a pentagonal surface, and Figure 2.26 shows a subdivision of a hexagonal surface.

2.3.4. *The Sierpinski triangle*

Several cellular decompositions can be proposed to represent the Sierpinski triangle. The two representations we present start from a main cell, the triangle itself (denoted by F). It is naturally divided into three parts, corresponding to the self-similarity property of the Sierpinski triangle (see Figure 2.27). A first solution consists of defining the minimum number of cells to connect these three subdivisions by vertices. For this purpose, we may simply define three incident vertices that will be used to define adjacency relations.

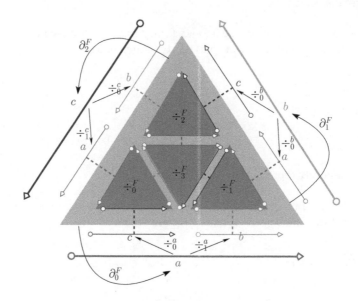

Figure 2.24. *Standard triangular subdivision, but with connections differing from the edges. For a color version of this figure, see www.iste.co.uk/gentil/geometric.zip*

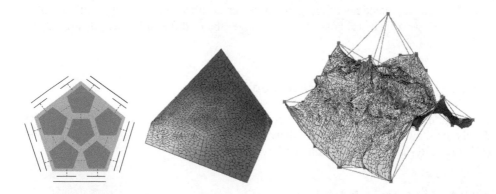

Figure 2.25. *Pentagonal subdivision. On the left, the incidence relations are symbolized using red dotted lines and adjacency relations using blue dotted lines. The face is divided into 6 parts and the edges into 2 parts; in the center, and on the right, respectively, visualization of the surface with a smooth geometry and a fractal geometry. For a color version of this figure, see www.iste.co.uk/gentil/geometric.zip*

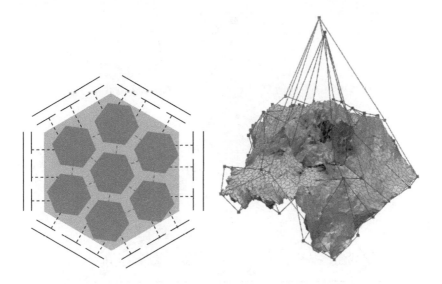

Figure 2.26. *On the left, incidence relations are symbolized using red dotted lines and adjacency relations using blue dotted lines. The face is divided into seven parts and the edges into three parts; on the right, the visualization of a hexagonal subdivision surface at iteration 4. The networks of control and subdivision points are in red and blue, respectively. For a color version of this figure, see www.iste.co.uk/gentil/geometric.zip*

The incidence constraints of the vertices S to the face F are then:

$$\div_0^F \partial_0^F \simeq \partial_0^F \div_0^S$$
$$\div_1^F \partial_1^F \simeq \partial_1^F \div_0^S$$
$$\div_2^F \partial_2^F \simeq \partial_2^F \div_0^S$$

and the adjacency constraints between the subdivided faces are:

$$\div_0^F \partial_1^F \simeq \div_1^F \partial_0^F$$
$$\div_1^F \partial_2^F \simeq \div_2^F \partial_1^F$$
$$\div_0^F \partial_2^F \simeq \div_2^{F'} \partial_0^{F'}$$

Another solution consists of defining edges incident to the Sierpinski triangle and then utilizing the vertices of the edges to achieve adjacency relations. This amounts to a standard triangular subdivision, for which the central triangle has been eliminated (see Figure 2.28).

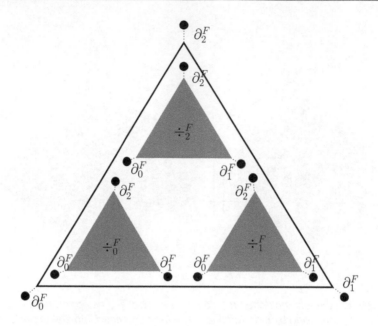

Figure 2.27. *Cellular decomposition and subdivision of the Sierpinski triangle.*
For a color version of this figure, see www.iste.co.uk/gentil/geometric.zip

For each edge $x = a$, b and c, the incidence and adjacency relations of a curve have to be written down:

$$\div_0^x \partial_0^x \simeq \partial_0^x \div_0^S$$

$$\div_1^x \partial_1^x \simeq \partial_1^x \div_0^S$$

$$\div_0^x \partial_1^x \simeq \div_1^x \partial_0^x$$

To this must be added the incidence relations of each edge to the face F:

$$\div_0^F \partial_0^F \simeq \partial_0^F \div_0^a$$

$$\div_1^F \partial_0^F \simeq \partial_0^F \div_1^a$$

$$\vdots$$

The adjacency relations between the subdivided faces are therefore useless, because they are induced by the incidence relations of the edges to the face. Given that the edges are continuous, it follows that the subdivided faces are adjacent to each other by means of the connecting vertices between the subdivided parts of the edges.

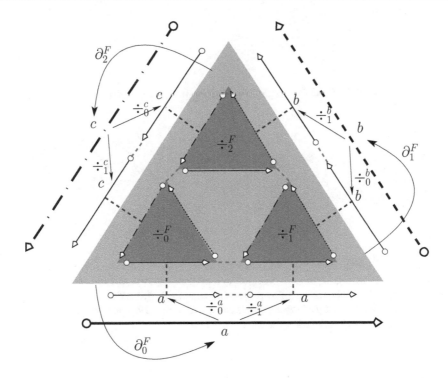

Figure 2.28. *Cellular decomposition and subdivision of the Sierpinski triangle with integration of edges bordering the triangle. Here, the vertices have not been represented. For a color version of this figure, see www.iste.co.uk/gentil/geometric.zip*

REMARK.– The second solution is more constrained than the first, in the sense that it is a special case. From a geometric point of view, by explicitly defining edge-type cells, we introduce a specific control by these cells. On the one hand, we choose the number of control points assigned to the edges and thus, influence them. As a result, it will be easier for us to influence their overall geometry, but we shall also be able to define their nature by specifying particular subdivision points to obtain, for example, Bezier curves, NURBS or "Takagi" curves. On the other hand, we can choose the number of control points influencing the face, without affecting the edges. Figure 2.29 shows an example of a Sierpinski triangle with Bezier curves as edges. The face has an internal dimension equal to 1, enabling the control of the internal geometry of the face without influencing the geometry of the edges: the edges remain Bezier curves without the Sierpinski triangle being included in a triangular Bezier patch. From a topological point of view, the fact that we have edge-type cells gives the possibility of connecting the Sierpinski triangle with another figure with a similar type of edge as incident cell.

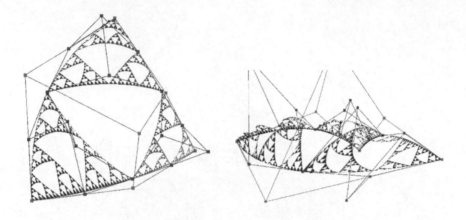

Figure 2.29. *Example of a Sierpinski triangle whose face, edges and vertices have an internal dimension equal to 1. The edges are Bezier curves. On the left, the top view highlights the Bezier curves as the edge. On the right, the side view shows that the Sierpinski triangle is not included in a Bezier patch. For a color version of this figure, see www.iste.co.uk/gentil/geometric.zip*

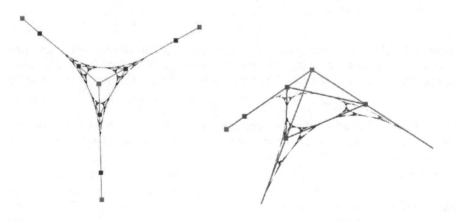

Figure 2.30. *Example of a Sierpinski triangle whose edges are uniform quadratic B-spline curves (on the left side, the view from above, on the right, the side view). For a color version of this figure, see www.iste.co.uk/gentil/geometric.zip*

COMMENTS ON FIGURE 2.30.– *Each vertex of every curve depends on two control points, which are shared by the common vertices of two adjacent curves. The three curves have a common control point located at the central part of the triangle. The Sierpinski triangle, on the other hand, depends on four control points (in red).*

2.3.5. *Penrose tilings*

As with the other examples, building a 2D Penrose tiling is relatively simple (Bandt and Gummelt 1997; Gelbrich 1997). However, a few precautions should be taken when describing the entire structure of the topological decomposition, if we want to be able to distort it at will. There are several types of Penrose tilings. We illustrate the tiling known as the "kite"-type (see Figure 2.31).

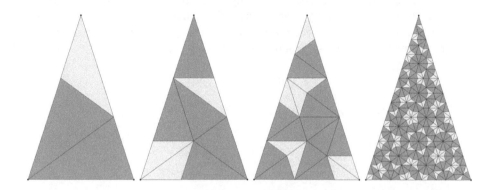

Figure 2.31. *Penrose tiling of the "kite" type at iterations 1, 2, 3 and 6. For a color version of this figure, see www.iste.co.uk/gentil/geometric.zip*

The difficulty is to identify the subdivision process of the faces and, especially, that of the edges, which is not explained as that of the faces. One will then merely have to represent them as incidence and adjacency constraints. The subdivision of the faces is presented by Figure 2.31 and Figure 2.32. The tiling consists of two types of tiles. The first tile is divided into two of the same type, and adding one of the other type. The other tile is divided into two tiles, one of each type. One should be careful about the orientation of the faces, as they are sometimes flipped. For the edges, a little more reflection is needed, since some edges are not subdivided at each stage. To express this property, we use the same process as for the square subdividing in two (see Figure 2.19). A complete solution to the subdivision system is presented in Figure 2.32. The edge a (in red) is divided into a single edge c (in green). The edge

c is divided into an edge e and an inverted edge c. Thereby, we obtain the following edge subdivision system:

$$c \longrightarrow e - c$$
$$a \longrightarrow c$$
$$b \longrightarrow a + c$$
$$e \longrightarrow -b$$

where the $+$ sign means that the edges follow one another and the $-$ sign means that the next edge is inverted. The edge and subdivision operators should then be named and the incidence and adjacency relations are written in accordance with the diagram. Figure 2.33 shows the result of a 3D embedding of the "kite"-type tiling, obtained after translation of the topological subdivision into a BC-IFS.

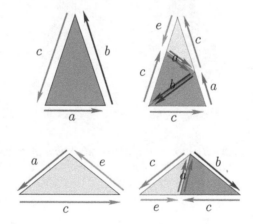

Figure 2.32. *Topological subdivision diagram of the faces and edges representing the Penrose tiling process of the "kite"-type. For a color version of this figure, see www.iste.co.uk/gentil/geometric.zip*

2.4. Volumes and lacunar objects

The design of volumes or volume structures follows exactly the same approach as that of surfaces, with an additional level in the cellular decomposition. Incidence relations need to be defined: from the vertices to the edges, from the edges to the faces and finally from the faces to the volumes. The adjacency relations must also respect the compatibility of orientations at every level (see section 4.1).

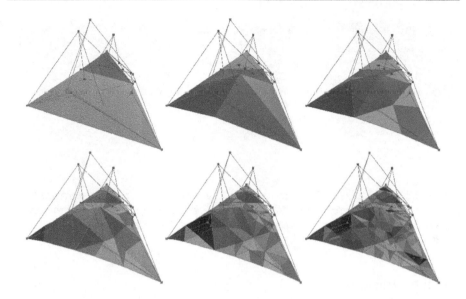

Figure 2.33. *Example of 3D surfaces constructed from the topological subdivision of the "kite"-type Penrose tiling. The surface has been creased using control points (in red) and subdivision points (in blue). For a color version of this figure, see www.iste.co.uk/gentil/geometric.zip*

The creation of the Menger sponge (see Figure 2.34) requires defining:

– a main cell (the Menger cube);

– bounded by six faces (Menger carpet);

– themselves bounded by four edges;

– each edge bounded by two vertices.

The main cell is subdivided into 20 parts of the same type, constrained by 24 adjacency relations. Each face is divided into eight parts of the same type, constrained by eight adjacency relations. Finally, each edge is subdivided into three, with two adjacency relations. To this we must add the adjacency relations on the incidence operators.

A simple solution to generate volume structures is to use the topological tensor product presented in section 4.2.1. Further examples are presented in sections 4.2 and 4.3.

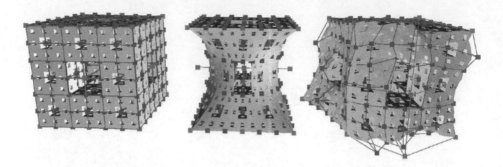

Figure 2.34. *Examples of the Menger sponge, represented with three iteration levels and defined by 27 control points. On the left, the classic Menger sponge; in the center, with a smooth deformation; on the right, with a fractal deformation. For a color version of this figure, see www.iste.co.uk/gentil/geometric.zip*

2.5. Tree structures

The design of the tree structure is very interesting, because it brings forward the structuring approach of the attractor in an incident cell to define the adjacency relations.

2.5.1. *Basic example*

Let us start with the example of a 2D tree structure, whose representation is given in Figure 2.35.

Tree structures are not strictly self-similar, in the sense that they are composed of at least two types of cells: the trunk, which cannot be subdivided (or, eventually, into itself), and branches, which are smaller copies of the tree. For our example, we symbolize the tree by a triangle and the trunk by a simple segment. The construction process starts with a tree A. It is divided into two trunks T and two trees A (see Figure 2.36). To define the connections between the subdivided elements of the tree, we introduce a vertex-type incident cell S and an edge-type incident cell B. The type that we attribute to these cells here is purely syntactic. The topological nature of these cells will only be truly defined by their own subdivision systems and their constraints of incidence and adjacency. On the one hand, we observe that the tree A is bounded by a set of cells of different topological kinds (a vertex and an edge) and, on the other hand, that these cells do not define the entire edge of the tree (some edges are not identified or specified). These unspecified edges will be dependent on subdivision operators, with no specific control (as any internal part of the attractor). For this example, we choose to subdivide the trunk into two identical parts, so as to build a curve. The automaton corresponding to these choices is presented in Figure 2.37.

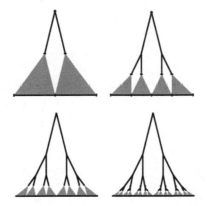

Figure 2.35. *Example of the construction of a 2D tree structure,
consisting of two branches. At each iteration, the tree is built by
adding branches to the ends of the branches of the previous stage*

Once the cellular structure is defined, we can explain the influence constraints.

– For the tree A:
   - the vertex S is incident to A:

$$\partial_1^A \div_0^S \simeq \div_0^A \partial_1^T \tag{2.24}$$

$$\partial_1^A \div_0^S \simeq \div_1^A \partial_1^T \tag{2.25}$$

   - the curve B is incident to A:

$$\partial_0^A \div_0^B \simeq \div_0^A \partial_0^A \tag{2.26}$$

$$\partial_0^A \div_1^B \simeq \div_1^A \partial_0^A \tag{2.27}$$

– For the trunk T:
   - the vertex S is incident to T:

$$\partial_0^T \div_0^S \simeq \div_0^T \partial_0^T \tag{2.28}$$

$$\partial_1^T \div_0^S \simeq \div_1^T \partial_1^T \tag{2.29}$$

Finally, the adjacency constraints connecting the subdivided elements for the tree are:

$$\div_2^A \partial_0^T \simeq \div_0^A \partial_1^A \tag{2.30}$$

$$\div_3^A \partial_0^T \simeq \div_1^A \partial_1^A \tag{2.31}$$

and, for the trunk:

$$\div_1^T \partial_0^T \simeq \div_0^T \partial_1^T \qquad\qquad\qquad [2.32]$$

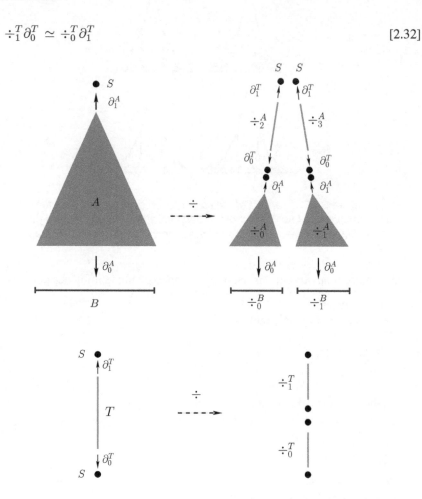

Figure 2.36. *Subdivision process of the tree structure of Figure 2.35 and cell decomposition. The tree (symbolized by a triangle) is bounded by an edge (at the bottom) and a vertex (on top). It is subdivided into two trunks and two trees. The trunk is divided into two trunks. The subdivision of the vertices has not been represented, for the sake of clarity. For a color version of this figure, see www.iste.co.uk/gentil/geometric.zip*

2.5.2. *Basic example with an undivided trunk*

It is possible to define the trunk by subdividing it into a single part (strictly equal to itself) and by imposing the identity operator as a subdivision operator (see

Figure 2.39). Since this is not contractive, the associated attractor depends on the primitive associated with the state of the trunk.

The automaton is then modified accordingly, as illustrated by Figure 2.41. It should be noted that, this time, the trunk must be bounded by three vertices in order to be able to achieve the connections between the subdivisions of the tree. We need to replace the incidence constraints of the vertices *vis-à-vis* the trunk with:

$$\partial_0^T \div_0^S \simeq \div_0^T \partial_0^T$$
$$\partial_1^T \div_0^S \simeq \div_0^T \partial_1^T$$
$$\partial_2^T \div_0^S \simeq \div_0^T \partial_0^T$$

and the adjacency contraints connecting the subdivided elements of the tree become :

$$\div_2^A \partial_0^T \simeq \div_0^A \partial_1^A$$
$$\div_2^A \partial_2^T \simeq \div_1^A \partial_1^A$$

and the adjacency constraints of the trunk disappear, since it is subdivided into only one copy.

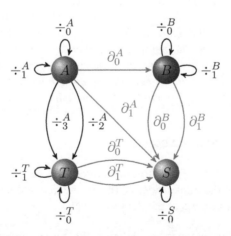

Figure 2.37. *Automaton describing the iterative construction process of the tree in Figure 2.35. For a color version of this figure, see www.iste.co.uk/gentil/geometric.zip*

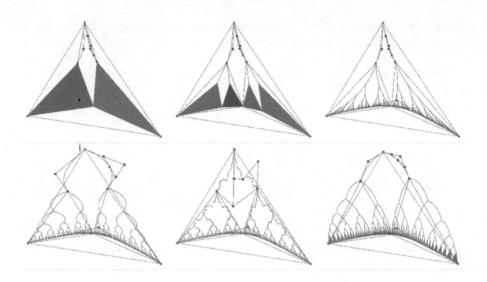

Figure 2.38. *Examples of projection of the tree's topological structure, defined by the automaton in Figure 2.37 and the incidence and adjacency constraints [2.14]–[2.32]. For a color version of this figure, see www.iste.co.uk/gentil/geometric.zip*

COMMENTS ON FIGURE 2.38.– *The top line shows iterations 1, 2 and 8. The subdivision matrices of the trunk are De Casteljau matrices, generating a quadratic Bezier curve. For the bottom line, the control points are identical to those of the figures on the upper line; the figure on the left is obtained by moving the subdivision points of the trunk; the middle figure is constructed from subdivisions of the trunk generating a fractal curve; for the figure on the right, the tree subdivisions have been modified with smaller contractions. For each of these figures, the tree leaf boundary, corresponding to the state B, is a quadratic Bezier curve.*

2.5.3. *Example by tiling the space*

Another possibility for defining tree structures consists of making a tiling of the space. The structure of the trunk, added to that of the tree, will always cover the same proportion of space, regardless of the iteration level.

In the 2D case, this amounts to defining the same subdivision system as the previous example, but adding an adjacency between the two subdivided trees by means of two new curve-type cells (which border the tree to the right and left). The tree subdivisions must themselves be adjacent to the trunk, by way of curves, equally bordering the top of the tree, the top and bottom of the trunk (as shown in Figure 2.42). This solution makes it easier to manage the risk of self-intersection of the structure, because of this cellular decomposition. The display primitive will have to satisfy the connection constraints to ensure the continuity of the structure.

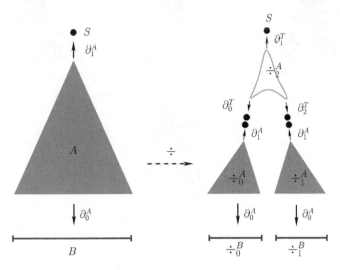

Figure 2.39. *Subdivision process of a tree structure whose trunk is subdivided into a single copy of itself. For a color version of this figure, see www.iste.co.uk/gentil/geometric.zip*

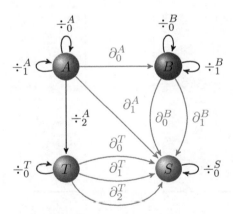

Figure 2.40. *Automaton describing the iterative construction process of the tree in Figure 2.35. For a color version of this figure, see www.iste.co.uk/gentil/geometric.zip*

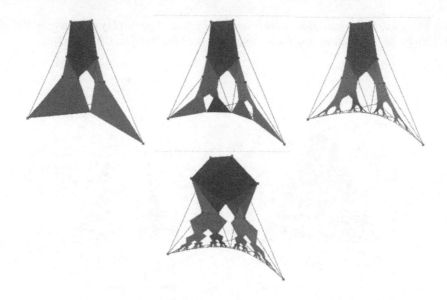

Figure 2.41. *Example of a tree structure whose trunk (green) is not subdivided (or subdivided into itself). The tree (red) is subdivided into a trunk and two trees. For a color version of this figure, see www.iste.co.uk/gentil/geometric.zip*

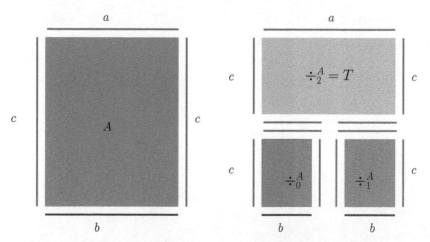

Figure 2.42. *Simplified representation of the incidence and adjacency relations of the tree tiling part of the 2D space. For a color version of this figure, see www.iste.co.uk/gentil/geometric.zip*

Figure 2.43 shows an example of a tree built on the principle of tiling a 3D space from a fractal hexagonal base.

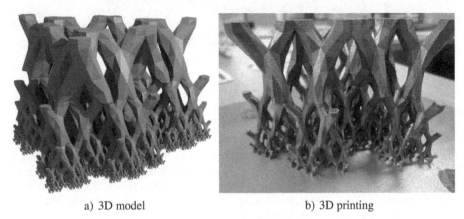

a) 3D model   b) 3D printing

Figure 2.43. *3D tree built on the principle of space tiling(source: project MODITERE no. ANR-09-COSI-014)*

2.6. Form assembly

The adjacency constraints can be used to achieve assemblies of shapes. The resolution of these constraints involves sharing:

– control points;

– adjacent edges;

– the columns corresponding to these control points in the subdivision matrices.

Figures 2.44, 2.45 and 2.46 give examples of assemblies of lacunar surface structures and surface structures with fractal geometry.

Figure 2.47 shows the assembly of 10 faces built from a Penrose tiling of the "kite"-type to obtain the configuration of the "sun".

Figure 2.48 shows an example of a density structure assembly from Menger sponges.

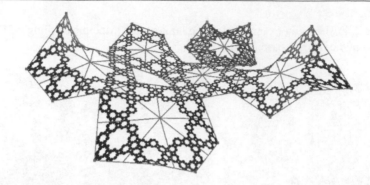

Figure 2.44. *Example of assembling fractal structures (built from an octagonal face). The control points are shown in red and the subdivision points of the assembled structure in blue. The assembly is carried out by adjacency relations that share control points of the basic structures, guaranteeing the C_0 continuity. For a color version of this figure, see www.iste.co.uk/gentil/geometric.zip*

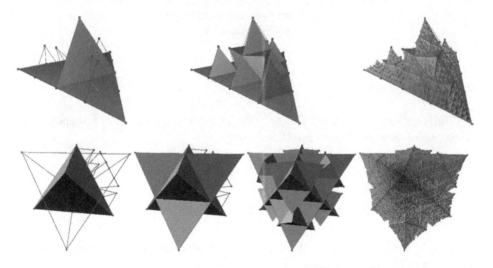

Figure 2.45. *Example of an assembly of triangular surface structures. The basic triangular face is subdivided into four, and a vertex has been added in place of the central face. For a color version of this figure, see www.iste.co.uk/gentil/geometric.zip*

COMMENTS ON FIGURE 2.45.– *Iterations 1, 2 and 5 are presented by the images of the first row. The assembly consists of constructing a closed surface by replacing each face of a tetrahedron by a basic fractal face. The images of the bottom row show the result for 0, 1, 2 and 6 iterations.*

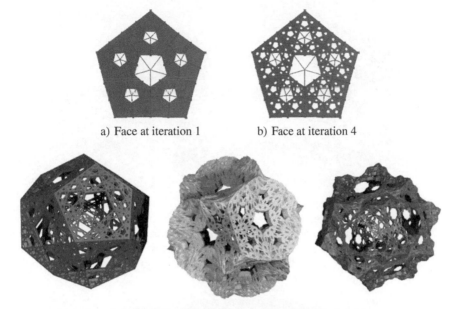

a) Face at iteration 1   b) Face at iteration 4

c) Assemblies of pentagonal faces in dodecahedron

d) 3D printing

Figure 2.46. *Examples of assemblies of pentagonal fractal faces following a dodecahedron. For a color version of this figure, see www.iste.co.uk/gentil/geometric.zip*

COMMENTS ON FIGURE 2.46.– *Figures 2.46(a) and (b) show the pentagonal face used to achieve the assembly at iteration levels 1 and 4, respectively. (c) represents the assembly with different changes in the face geometry. The adjacency constraints between the faces ensure the consistency of the geometry of the dodecahedron. (d) shows the 3D printing of one of the assemblies.*

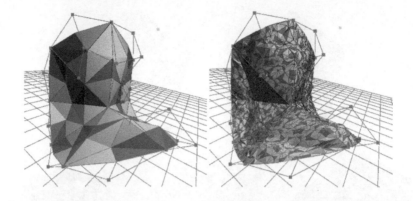

Figure 2.47. *Assembly of 10 3D Penrose tilings of the "kite"-type to form the configuration of the "sun". The surfaces are visualized at iterations 2 and 6. For a color version of this figure, see www.iste.co.uk/gentil/geometric.zip*

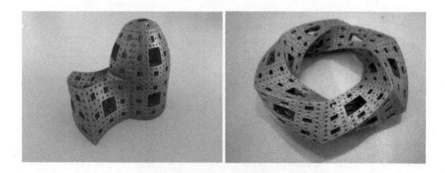

Figure 2.48. *Examples of assemblies built from Menger sponges and manufactured by direct laser metal sintering. On the left, the structure is composed of three deformed Menger sponges. On the right, six Menger sponges constitute the torus. A twist has been applied resembling a Möbius strip. The parts were designed by the LIB and manufactured by IPC (source: project MODITERE no. ANR-09-COSI-014). For a color version of this figure, see www.iste.co.uk/gentil/geometric.zip*

3

Surface NURBS, Subdivision Surfaces and BC-IFS

Non-uniform rational B-splines (NURBS) play an essential role in computer-aided design (CAD). These geometric representations are used to describe polynomial and rational surfaces using a grid of control points. The designer can then change the shape of the surface by simply moving the control points. For savvy users, other parameters can be manipulated. The weights associated with every control point define the degree of influence of the point on the surface. Other parameters, such as the grid of knots (a knot represents a numerical value), make it possible to change the basis functions of the model. The latter accurately define how each control point influences the surface, depending on the parameters (u,v). The values of these knots are generally adjusted to obtain certain mathematical properties of the resulting surfaces (differentiabillity, approximation, interpolation, etc.). They can also be adjusted automatically by means of optimization algorithms. The mathematical properties of NURBS are clearly established, making it the preferred representation for CAD.

In the context of this book, we are focusing on NURBS for two reasons. The first reason is related to the interoperability of software programs and the compatibility of their representation models. We are faced with this problem when employing fractal structures within an industrial context, for example to design lacunar, lightened structures. These structures will have to be integrated into an existing mechanical system. Fractal shapes should be adjusted to forms originating from standard CAD, usually described using NURBS. The second reason is related to the properties of the NURBS. We have known for a long time that Bezier curves and surfaces (respectively, B-splines), from which NURBS originate, can be built iteratively using the De Casteljau algorithm (Casteljau 1985) (respectively, the Chaikin algorithm

---

For a color version of all the figures in this chapter, see www.iste.co.uk/gentil/geometric.zip

(Chaikin 1974; Riesenfeld 1975)). We shall see that this process is general and can be extended to NURBS. NURBS can then be defined by a C-IFS and integrated as attractor sub-cells.

Subdivision surfaces are more commonly used for the production of animated films, for which geometric constraints are generally less important. They have the advantage of being able to generate smooth surfaces based on a network of arbitrary control points, not necessarily regular as in the case of NURBS. They are defined from iterative computational processes that are naturally represented by automatons. Therefore, subdivision surfaces can be described by the formalism of C-IFS.

3.1. Bezier curves and surfaces

Bezier curves are polynomial curves with poles, namely defined from a set of control points and basis functions. We recall that a Bezier curve is given by:

$$C(t) = \Sigma_{i=0}^{m-1} P_i B_i^{m-1}(t) \tag{3.1}$$

where $t \in [0, 1]$, $\{P_0, \cdots, P_{m-1}\}$ is the set of the m control points (P_i is a row vector formed of the coordinates of the ith point, and B_i^{m-1} are Bernstein polynomials).

Control points are points of the space in which we want to define a curve: usually $\mathbb{R}^2$ or $\mathbb{R}^3$ but they can be defined in $\mathbb{R}^n$, $n \geq 4$. In order for a curve to be defined according to equation [3.1], the basis functions must verify: $\forall t \in [0, 1]$, $\Sigma_{i=0}^{m-1} B_i^{m-1}(t) = 1$. This property guarantees that $\Sigma_{i=0}^{m-1} P_i B_i^{m-1}(t)$ is indeed a barycentric combination, therefore the result is independent of the frame of reference in which the coordinates of control points are expressed. We can also write equation [3.1] in the following vector form:

$$C(t) = P B_i^{m-1}(t) \tag{3.2}$$

where $P = \begin{pmatrix} P_0 & \cdots & P_{m-1} \end{pmatrix}$ and $B_i^{m-1}(t) = \begin{pmatrix} B_0(t) \\ \vdots \\ B_{m-1}(t) \end{pmatrix}$.

The De Casteljau algorithm shows that a Bezier curve of control points $P = \begin{pmatrix} P_0 & \cdots & P_{m-1} \end{pmatrix}$ can be subdivided into two other Bezier curves, with respective control points $Q_0 = \begin{pmatrix} q_0 & \cdots & q_{m-1} \end{pmatrix}$ and $Q_1 = \begin{pmatrix} q_{m-1} & \cdots & q_{2m-2} \end{pmatrix}$, such that the first curve corresponds to the first portion of the initial curve (for example $t \in [0, \frac{1}{2}]$) and the second curve corresponds to the second part (for example, for $t \in [\frac{1}{2}, 1]$, as illustrated by Figure 3.1). The computation of control point vectors Q_0 and Q_1 is achieved by the De Casteljau matrices D_0 and D_1: $Q_0 = PD_0$ and $Q_1 = PD_1$. Therefore, we have the following property:

$$C([0,1]) = PB([0,1])$$
$$= Q_0B([0,1]) \cup Q_1B([0,1])$$
$$= PD_0B([0,1]) \cup PD_1B([0,1])$$

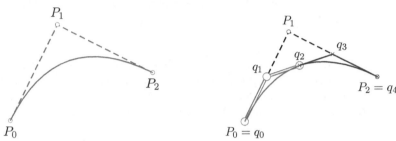

Figure 3.1. *On the left is an example of a quadratic Bezier curve, defined by three control points P_0, P_1 and P_2*

COMMENTS ON FIGURE 3.1.– *On the right-hand side, this curve can be divided into two quadratic Bezier curves, with the respective control points q_0, q_1 and q_2 for the curve in red and q_2, q_3 and q_4 for the curve in blue. These control points can be calculated using the De Casteljau matrices from the control points of the initial curve:* $(q_0 \; q_1 \; q_2) = (P_0 \; P_1 \; P_2) D_0$ *and* $(q_2 \; q_3 \; q_4) = (P_0 \; P_1 \; P_2) D_1.$

With this property being true for any vector of control points P, we infer that:

$$B([0,1]) = D_0B([0,1]) \cup D_1B([0,1]) \qquad\qquad [3.3]$$

Each column of the De Casteljau matrices corresponds to the coefficients of a barycentric combination of control points. Thereby, the sum of their coefficients is equal to 1. In addition, the transformations represented by these matrices are contractive (an eigenvalue is equal to 1 and the others are of a modulus less than 1). We deduce that $B([0,1])$ is the unique attractor of the IFS $\{D_0, D_1\}$. This attractor is defined in the barycentric space of dimension m that we denote by BI^m. $\mathcal{A}(\{D_0, D_1\})$ is the curve $B(t), t \in [0,1]$ defined by the Bernstein polynomials of dimension m. The curve $C([0,1])$ is the projection of $B(t)$, following the control points $P = (P_0 \; \cdots \; P_{m-1})$. Any Bezier curve can thus be represented by the C-IFS in Figure 3.2.

For a quadratic Bezier curve, $m = 3$, $P = (P_0 \; P_1 \; P_2)$;
$$D_0 = \begin{pmatrix} 1 & 1/2 & 1/4 \\ 0 & 1/2 & 1/2 \\ 0 & 0 & 1/4 \end{pmatrix} \text{ and } D_1 = \begin{pmatrix} 1/4 & 0 & 0 \\ 1/2 & 1/2 & 0 \\ 1/4 & 1/2 & 1 \end{pmatrix}.$$

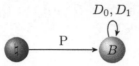

Figure 3.2. *C-IFS automaton whose attractor is a Bezier curve. This automaton simply consists of a state B with two transitions (to which the De Casteljau matrices (D_0 and D_1) are associated) and a projection into the modeling space (following the matrix of the control points P)*

Specifically, the De Casteljau algorithm builds an IFS morphism (see section 1.1.9) between $[0, 1]$ and $B(t)$. The first half of the Bezier curve is $B([0, \frac{1}{2}]) = B(\tau_0[0, 1])$, where τ_0 is the transformation defined by $\forall x \in \mathbb{R}, \tau_0(x) = \frac{x}{2}$. We have seen that $B(\tau_0 t) = D_0 B(t), t \in [0, 1]$. By analogy for the second part of the curve, we have: $B(\tau_1 t) = D_1 B(t), t \in [0, 1]$ where $\forall x \in \mathbb{R}, \tau_1(x) = \frac{x}{2} + \frac{1}{2}$. The only multiple address point of $\mathcal{A}(\{\tau_0, \tau_1\})$ is the point $\frac{1}{2} = \phi(01^\omega) = \phi(10^\omega)$, where ϕ is the addressing function associated with the IFS $\{\tau_0, \tau_1\}$. By denoting Φ, the addressing function of $\{D_0, D_1\}$, we have:

$$\Phi(1^\omega) = \begin{pmatrix} 0 \\ 0 \\ 1 \end{pmatrix} = c_1 \text{ fixed point of } D_1$$

$$\Phi(0^\omega) = \begin{pmatrix} 1 \\ 0 \\ 0 \end{pmatrix} = c_0 \text{ fixed point of } D_0$$

$$\Phi(01^\omega) = D_0 c_1 = \begin{pmatrix} 1/4 \\ 1/2 \\ 1/4 \end{pmatrix}$$

$$\Phi(10^\omega) = D_1 c_0 = \begin{pmatrix} 1/4 \\ 1/2 \\ 1/4 \end{pmatrix}$$

Therefore, $\Phi(01^\omega) = \Phi(10^\omega)$, and as such B is indeed an IFS morphism, carrying the parameter space $[0, 1]$ over onto the Bezier curve. These properties, highlighted here on quadratic curves, hold for every degree.

3.2. Uniform B-spline curves and surfaces

For uniform B-spline curves, the reasoning is strictly identical to that of the Bezier curves, but employing the Chaikin algorithm (see Figure 3.3). The difference lies in the dimensions of the vertices of the B-spline curve of degree d, since these depend on d control points (instead of 1 for the vertices of the Bezier curves).

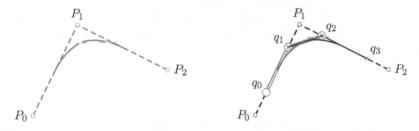

Figure 3.3. *Illustration of the self-similarity property of uniform quadratic B-spline curves, using the Chaikin algorithm. On the left is a uniform quadratic B-spline curve defined from three control points P_0, P_1 and P_2*

COMMENTS ON FIGURE 3.3.– *On the right, the curve is divided into two quadratic B-spline curves. The first part in red is defined from the control points q_0, q_1 and q_2. The second part in blue is defined from the control points q_1, q_2 and q_3. The control points of the subdivided curves are calculated from the control points of the initial curve, using the Chaikin matrices:* $C_0 = \begin{pmatrix} 3/4 & 1/4 & 0 \\ 1/4 & 3/4 & 3/4 \\ 0 & 0 & 1/4 \end{pmatrix}$ *and* $C_1 \begin{pmatrix} 1/4 & 0 & 0 \\ 3/4 & 3/4 & 1/4 \\ 0 & 1/4 & 3/4 \end{pmatrix}$:
$(q_0\ q_1\ q_2) = (P_0\ P_1\ P_2)C_0$ *and* $(q_1\ q_2\ q_3) = (P_0\ P_1\ P_2)C_1$.

We can see that for a uniform quadratic B-spline curve, the vertices depend on two control points. For its part, the curve depends on three control points. It is divided into two parts and the Chaikin matrices D_0 and D_1 are subdivision operators. The IFS $\{D_0, D_1\}$ has as attractor, the basis functions $N(t) = \begin{pmatrix} N_0(t) \\ N_1(t) \\ N_2(t) \end{pmatrix}$ of the uniform quadratic B-spline curves, the curve is then defined from the control points $P = (P_0\ P_1\ P_2)$ by: $C(t) = PN(t)$. The associated automaton is then the same as for the Bezier curves (see Figure 3.2), by replacing the De Casteljau matrices with the Chaikin matrices. We do not have to apply the incidence and adjacency constraints, since the subdivision matrices are determined by the properties of the basis functions defining curves. We find the structures inducing the connections between the two subdivisions of the attractor, in the subdivision matrices (see section 1.3.5): the last two columns of C_0 are identical to the first two columns of C_1.

3.3. Generalization

This approach can be generalized to most curves used in CAD. First of all, we present polynomial curves of any degree. Then, we address rational curves. Finally, we detail the non-uniform case of NURBS curves. The transition to surfaces is immediate by tensor product.

3.3.1. *Polynomial curves*

The properties that we have highlighted for Bezier curves and B-spline curves hold for any polynomial curve defined by:

$$C(t) = PB(t)$$

where $P = (P_0 \quad \cdots \quad P_{m-1})$ represents a vector of control points and $B(t) = \begin{pmatrix} B_0(t) \\ \vdots \\ B_{m-1}(t) \end{pmatrix}$ is a basis of polynomial functions of degree $m-1$, defined on $[0,1]$ and with values in the barycentric space of dimension m (BI^m), that is, verifying $\forall t \in [0,1], \Sigma_{i=0}^{m-1} B_i(t) = 1$.

Consider a subdivision of the parameter space $[0,1]$ into sub-intervals $[t_i, t_{i+1}]$, $i \in \{0, \cdots, m-2\}$ with $t_i < t_j$ if $i < j$. This subdivision may be represented by the IFS $\{\tau_i\}_{i \in \{0, \cdots, m-2\}}$ with $\tau_i(t) = \frac{t}{(t_{i+1}-t_i)} + t_i$. τ_i transforming $[0,1]$ into the interval $[t_i, t_{i+1}]$. Consider the variable change $X = \tau_i x$:

$$C(\tau_i t) = PB(\tau_i t) \tag{3.4}$$

$B(t)$ is a basis of polynomial functions of degree $d = m-1$, and $B(\tau_i t)$ is also one and can be expressed in the basis $B(t)$, that is:

$$\forall i \in \{0, \cdots, m-2\}, B(\tau_i t) = T_i B(t)$$

where T_i is the change of basis matrix. Here, we do not have to worry about the connection conditions since $B(t)$ is given and continuous by definition. We deduce that:

$$B([0,1]) = B(\bigcup_{i=0}^{m-2} \tau_i[0,1])$$

$$= \bigcup_{i=0}^{m-2} B(\tau_i[0,1])$$

$$= \bigcup_{i=0}^{m-2} T_i B([0,1])$$

$B([0,1])$ is indeed self-similar with respect to the transformations T_i representing the change of basis between $B(\tau_i t)$ and $B(t)$. Therefore, the IFS $\mathbb{T} = \{T_i\}_{i \in \{0, \cdots, m-2\}}$ has as attractor $B(T), t \in [0,1]$.

3.3.2. *Rational curves*

The transfer of the set of results to rational curves is immediate. One simply has to build the curves from the same representations in the barycentric space to make a first projection in a homogeneous space (using the control points, with an additional coordinate w representing the associated weight), then compute the projection onto the modeling space (by dividing each coordinate by the weight). A rational curve is given by:

$$C(t) = \frac{\Sigma_{i=0}^{m-1} P_i w_i B_i(t)}{\Sigma_{i=0}^{m-1} w_i B_i(t)}$$

where w_i represents the ith weight assigned to the ith control point. A similar formulation is then:

$$\tilde{C}(t) = \tilde{P} B(t)$$

with $P = (\tilde{P}_0, \cdots, \tilde{P}_{m-1})$ and $\tilde{P}_i = \begin{pmatrix} P_{i|x} \\ P_{i|y} \\ P_{i|z} \\ P_{i|w} \end{pmatrix}$.

We then have $C(t) := \begin{pmatrix} \frac{\tilde{C}_{i|x}(t)}{\tilde{C}_{i|w}(t)} \\ \frac{\tilde{C}_{i|y}(t)}{\tilde{C}_{i|w}(t)} \\ \frac{\tilde{C}_{i|z}(t)}{\tilde{C}_{i|w}(t)} \end{pmatrix}$.

3.4. NURBS curves

NURBS are an extension of rational B-splines. They are also defined from m weighted control points $C(t) = \frac{\Sigma_{i=0}^{m-1} P_i w_i N_{i,d}(t)}{\Sigma_{i=0}^{m-1} w_i N_{i,d}(t)}$, but they are enriched with a *knot vector*, which allows us to modify the basis functions and thus the areas of influence of the control points. This knot vector is an increasing sequence of $l = m + d + 1$ real values $(t_0, t_1, \ldots, t_{l-1})$. The basis functions $N_{i,d}(t), i \in \{0, m - 1\}$ are polynomial functions of degree d, defined on the respective intervals $t_i < t < t_{i+1}$. For every t, they describe a partition of the unit. They can be built in different ways. Here, we recall the recursive Cox–de Boor formula (Cox 1972):

$$N_{i,0} = \begin{cases} 1 \text{ if } t_i \leq t \leq t_{i+1} \\ 0 \text{ otherwise.} \end{cases}$$

$$N_{i,d}(t) = \frac{t - t_i}{t_{i+d} - t_i} N_{i,d-1}(t) + \frac{t_{i+d+1} - t}{t_{i+d+1} - t_{i+1}} N_{i+1,d-1}(t)$$

with $\frac{0}{0} = 0$ as convention.

Every function $N_{i,d}$ is zero outside of the interval $[t_i, t_{i+d+1}]$, and positive or zero in $[t_i, t_{i+d+1}]$. Figure 3.4 shows the graph of these functions for $d = 2$. The curve is built piecewise and is composed of $n = m - d$ parts. Each part is defined for the parameter $t \in u_i = [t_i, t_{i+1}]$ and the curve for $t \in [t_d, t_m]$. When examining the Cox–De Boor formula, we note that for the evaluation of the basis functions on $[t_d, t_m]$, the first and last intervals (u_0 and u_{m+d-1}) are not necessary. Therefore, for the calculation of basis functions, we consider that the intervals u_i are strictly necessary, that is, $u_1, \cdots, u_{m+d-2}$.

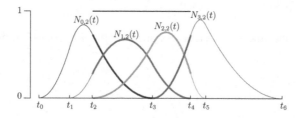

Figure 3.4. *Blending B-spline functions of the second degree. Their support is of three consecutive internodal intervals*

EXAMPLE.– For a NURBS curve of degree $d = 2$, we choose $m = 4$ control points P_0, P_1, P_2, P_3. The number of knots is then $l = m + 1 + 2 = 7$ and the knot vector is $(t_0, \cdots, t_6)$. The intervals for which the curve is defined are u_2 and u_3. The set of knots strictly necessary to evaluate the basis functions is $(t_1, \cdots, t_5)$ corresponding to the four knot intervals u_1, u_2, u_3 and u_4.

3.4.1. *Knot-doubling*

To understand the system of subdivision of NURBS curves, we rely on knot-doubling (Sederberg *et al.* 1998). This technique shows that it is possible to calculate a set of new control points $Q = (q_0, \cdots, q_{2m-d-1})$ associated with a new knot vector $(t_0, \frac{t_1-t_0}{2}, t_1, \ldots, \frac{t_{l-1}-t_{l-2}}{2}, t_{l-1})$, and representing the same NURBS curve. Considering only useful intervals, the knot interval vector is (u_1, u_2, u_3, u_4), which becomes $(\frac{u_1}{2}, \frac{u_1}{2}, \frac{u_2}{2}, \frac{u_2}{2}, \frac{u_3}{2}, \frac{u_3}{2}, \frac{u_4}{2}, \frac{u_4}{2})$ by duplication. If we only retain the useful intervals (for example, by getting rid of both ends), we get: $(\frac{u_1}{2}, \frac{u_2}{2}, \frac{u_2}{2}, \frac{u_3}{2}, \frac{u_3}{2}, \frac{u_4}{2})$. This knot interval vector can be made more readable by multiplying every value by 2: $(u_1, u_2, u_2, u_3, u_3, u_4)$. This does not change the basis functions since what is important are the ratios between these internodal values. Figure 3.5 shows the effect of doubling knots. The curve is built in two parts (one red and one green). Similarly to the Chaikin algorithm, the red part (on top of the figure) was broken down into two parts, red and green (at the bottom of the figure), and the green part (top of the figure) into two parts, blue and purple, respectively (at the bottom of the figure) (Morlet *et al.* 2019).

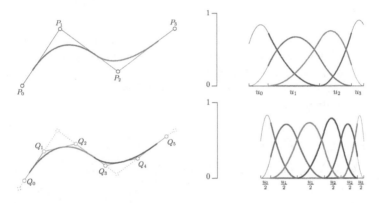

Figure 3.5. *The control points (P_0, P_1, P_2, P_3) and the knot vector (u_0, u_1, u_2, u_3) exactly defined the same NURBS curve as the control points ($Q_0, \ldots, Q_5$) with the knot vector ($\frac{u_1}{2}, \frac{u_2}{2}, \frac{u_2}{2}, \frac{u_3}{2}, \frac{u_3}{2}, \frac{u_4}{2}$)*

For curves of degree 2, the computation of new control points is done by using the following rules:

$$Q_{2i} = \frac{(u_i + 2u_{i+1})P_i + u_i P_{i+1}}{2(u_i + u_{i+1})} \qquad [3.5]$$

$$Q_{2i+1} = \frac{u_{i+1}P_i + (2u_i + u_{i+1})P_{i+1}}{2(u_i + u_{i+1})} \qquad [3.6]$$

Knot-doubling allows us to break down each part of the curve into two subparts. We will study this decomposition in order to deduce an iterative generation system, first for curves of degree 2, then for the curves degree 3, and finally, we will propose a generalization.

3.4.2. *Curve subdivision*

Any NURBS curve with degree d, defined by m control points $m \geq d + 1$ and an knot interval vector $u_1, \cdots, u_{m+d-2}$, can be seen as $n = m - d$ curves defined by $d + 1$ control points and $2d - 1$ internodal values. We will show how, for each part, it is possible to associate a C-IFS.

3.4.2.1. *Second degree NURBS*

The construction of the second-degree curve in Figure 3.5, defined from the four control points (P_0, P_1, P_2, P_3) and the knot interval vector (u_1, u_2, u_3, u_4), can be achieved with two second-degree curves. For the first curve (shown in red in the figure), the control points are (P_0, P_1, P_2) and the knot interval vector is (u_1, u_2, u_3). The second is obtained with the control points (P_1, P_2, P_3) and the knot interval vector (u_2, u_3, u_4).

Consider the first part of the curve (in red). The knot-doubling algorithm decomposes it into two subparts, respectively, defined by the control points (q_0, q_1, q_2) and (q_1, q_2, q_3) and the knot interval vectors (u_1, u_2, u_2) and (u_2, u_2, u_3). As we have seen, the control points are calculated from the formulas [3.5] and [3.6], which can be written in matrix form by:

$$(q_0, q_1, q_2) = (P_0, P_1, P_2)T_0(u_1, u_2, u_3) \tag{3.7}$$

$$(q_1, q_2, q_3) = (P_0, P_1, P_2)T_1(u_1, u_2, u_3) \tag{3.8}$$

$$T_0(u, v, w) = \begin{pmatrix} \frac{u+2v}{2(u+v)} & \frac{v}{2(u+v)} & 0 \\ \frac{u}{2(u+v)} & \frac{2u+v}{2(u+v)} & \frac{v+2w}{2(v+w)} \\ 0 & 0 & \frac{v}{2(v+w)} \end{pmatrix} \tag{3.9}$$

$$T_1(u, v, w) = \begin{pmatrix} \frac{v}{2(u+v)} & 0 & 0 \\ \frac{2u+v}{2(u+v)} & \frac{v+2w}{2(v+w)} & \frac{w}{2(v+w)} \\ 0 & \frac{v}{2(v+w)} & \frac{2v+w}{2(v+w)} \end{pmatrix} \tag{3.10}$$

Therefore, overall we find some kind of self-similarity of the curve, in the sense that $Pf(\bullet) = PT_0(u_1, u_2, u_3)f'(\bullet) \cup PT_1(u_1, u_2, u_3)f''(\bullet)$. However, the basis functions of the new curves ($f'(\bullet)$ and $f''(\bullet)$) are not quite identical to those of the original curve ($f(\bullet)$). As a matter of fact, these basis functions are defined by the knot interval vectors (u_1, u_2, u_2) for $f'(\bullet)$ and (u_2, u_2, u_3) for $f''(\bullet)$.

If we reapply knot-doubling on these new curves, for the first two sub-curves, such that from the control points (q_0, q_1, q_2), we obtain four new control points

(R_0, R_1, R_2, R_3) with (u_1, u_2, u_2, u_2) as the knot interval vector. We can then decompose this portion of the curve into two: a first curve defined by the control points (R_0, R_1, R_2) and the knot interval vector (u_1, u_2, u_2) and a second curve defined by the control points (R_1, R_2, R_3) and the knot interval vector (u_2, u_2, u_2). We then note that the first curve has the same knot interval vector as the initial curve, that is to say, the same basis functions. In addition, the second has a uniform knot interval vector, in other words it is a uniform B-spline curve:

$$(R_0, R_1, R_2) = (Q_0, Q_1, Q_2)T_0(u_1, u_2, u_2)$$
$$(R_1, R_2, R_3) = (Q_0, Q_1, Q_2)T_1(u_2, u_2, u_2)$$

If we continue to duplicate the knots, for the part of the curve knot interval vector (u_1, u_2, u_2) we will get a curve with an identical knot interval vector (u_1, u_2, u_2) and a second curve with a uniform knot interval vector. For a uniform part, knot-doubling will always give two uniform parts.

Let us reconsider the part of the curve of knot interval vector (u_2, u_2, u_3). Similarly and symmetrically, the decomposition is achieved in a curve of uniform knot interval vector (u_2, u_2, u_2) and a curve of knot interval vector (u_2, u_2, u_3):

$$(R_2, R_3, R_4) = (Q_1, Q_2, Q_3)T_0(u_2, u_2, u_3)$$
$$(R_3, R_4, R_5) = (Q_1, Q_2, Q_3)T_1(u_2, u_2, u_2)$$

This subdivision process is illustrated in Figure 3.6.

We can now describe the subdivision process using an automaton, in which the names assigned to the states identify the type of knot interval vector $((uvw), (uvv), (vvw), (vvv))$. Each letter represents a distinct knot interval value (see Figure 3.7).

Therefore, the automaton clearly shows the dynamics of the subdivision process. A curve with a non-uniform knot interval vector uvw is decomposed into two curves of the uvv and vvw type. The subdivision of uvv gives a curve uvv and a uniform curve vvv. There will still be a non-uniform part of the curve uvv to the left of the curve, which will be increasingly smaller. The rest is made up of uniform parts. In a symmetrical way, the same phenomenon occurs for the right-hand side of the curve of the vvw type. Figure 3.8 shows the portions of non-uniform curves according to the subdivision level in red. In this illustration, it must be understood that the parts become uniform under the refinement of the control points.

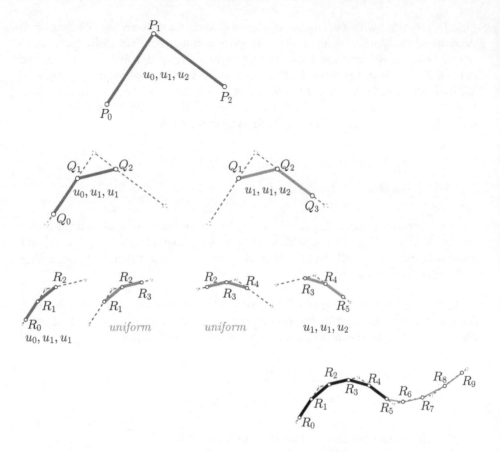

Figure 3.6. *Subdivision scheme obtained by duplication of knots*

3.4.2.2. NURBS of degree 3

The procedure applied to curves of degree 2 applies identically to those of degree 3.

Consider the minimum configuration to build such a curve: four control points (P_0, P_1, P_2, P_3) and a knot interval vector $(u_0, u_1, u_2, u_3, u_4)$.

The knot-doubling algorithm produces five new control points $(Q_0, Q_1, Q_2, Q_3, Q_4)$ and a vector of six useful knot intervals $(u_1, u_1, u_2, u_2, u_3, u_3)$. This configuration represents two parts of the curve defined by (Q_0, Q_1, Q_2, Q_3) and $(u_1, u_1, u_2, u_2, u_3)$ for the first and (Q_1, Q_2, Q_3, Q_4) and $(u_1, u_2, u_2, u_3, u_3)$ for the second. We find that the process is identical to that of second-degree curves. However, given that knot interval vectors have more components, the number of the types of knot interval vectors produced by the algorithm will be greater. By continuing this process until we find already known types of knot interval vectors, we can automatically build the C-IFS automaton representing the subdivision process (as shown in Figure 3.9).

$$T_0(u, v, w, x, y) = \begin{pmatrix} a & b & 0 & 0 \\ 1-a & c & a' & b' \\ 0 & 1-b-c & 1-a' & c' \\ 0 & 0 & 0 & 1-b'-c' \end{pmatrix} \qquad [3.11]$$

$$T_1(u, v, w, x, y) = \begin{pmatrix} b & 0 & 0 & 0 \\ c & a' & b' & 0 \\ 1-b-c & 1-a' & c' & a'' \\ 0 & 0 & 1-b'-c' & 1-a'' \end{pmatrix} \qquad [3.12]$$

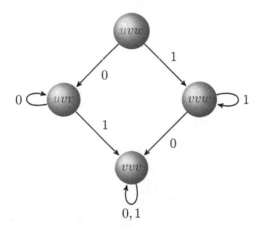

Figure 3.7. *Automaton of the C-IFS representing the iterative process of a second-degree NURBS curve. With each transition δ(uvw, 0) is associated the subdivision matrix $T_0(u, v, w)$ (see equation [3.9]) and with δ(uvw, 0) the matrix $T_1(u, v, w)$ (see equation [3.10])*

Figure 3.8. *From left to right: The non-uniform quadratic curve defined from four initial control points (namely, two minimal curve portions of three control points), the curve at iteration level 2 with refined control points, the curve at iteration level 5*

COMMENTS ON FIGURE 3.8.– *The parts in red represent portions of a curve of non-uniform knot interval vectors, and the parts in green represent portions of a uniform curve.*

with:

$$a = \frac{v + 2w}{2(u + v + w)}$$

$$1 - a = \frac{2u + v}{2(u + v + w)}$$

$$b = \frac{w}{2(v + w)} \frac{v + 2w}{2(u + v + w)}$$

$$c = \frac{w}{2(v + w)} \frac{2u + v}{2(u + v + w)} + \frac{2v + w}{2(v + w)} \frac{v + 2w + 2x}{2(v + w + x)}$$

$$1 - b - c = \frac{2v + w}{2(v + w)} \frac{v}{2(v + w + x)}$$

$$a' = \frac{w + 2x}{2(v + w + x)}$$

$$1 - a' = \frac{2v + w}{2(v + w + x)}$$

$$b' = \frac{x}{2(w + x)} \frac{w + 2x}{2(v + w + x)}$$

$$c' = \frac{x}{2(w + x)} \frac{2v + w}{2(v + w + x)} + \frac{2w + x}{2(w + x)} \frac{w + 2x + 2y}{2(w + x + y)}$$

$$1 - b' - c' = \frac{2w + x}{2(w + x)} \frac{w}{2(w + x + y)}$$

$$a'' = \frac{x + 2y}{2(w + x + y)}$$

$$1 - a'' = \frac{2w + x}{2(w + x + y)}$$

3.4.2.3. *NURBS of any degree*

To generalize this construction to a NURBS curve of any degree, we need to automate, on one hand, the generation of the automaton and, on the other hand, the calculation of subdivision matrix coefficients T_0 and T_1.

The building process of the automaton has almost been described for curves of degree 3. Consider a curve of degree d defined for the minimum number of control points needed for its definition, for example, $m = d + 1$. To calculate the basis functions of such a curve, it is necessary to have $2d - 1$ knot interval values. If we number these values from 0 to $2d - 2$, the central knot interval u_{d-1} represents the interval on which the curve is defined. The knot-doubling algorithm will produce $m + 1$ new control points grouped into two sets of m control points $(Q_0, \cdots, Q_{m-1})$ and $(Q_1, \cdots, Q_m)$ (one for each sub-curve). The knot interval vectors associated with each of these curves comprise $2d - 1$ values that we can simply deduce. Each of the knot intervals is split in half. The refined curve is always defined on the interval of parameters corresponding to the interval u_{d-1}, which, after duplication, is represented by two intervals of length $\frac{u_{d-1}}{2}$. Thereby, the first sub-curve, defined by the control points $(Q_0, \cdots, Q_{m-1})$, is obtained by a knot interval vector centered on the first split interval $\frac{u_{d-1}}{2}$, which has the $d - 1$ subdivided intervals attached on each side (to form a vector of $2d - 1$ knot interval values). After splitting the intervals, the first $d - 1$ and the last $d - 1$ are not useful for the construction of basis functions. Figure 3.10 illustrates these properties with an example of knot-doubling (for a third-degree curve).

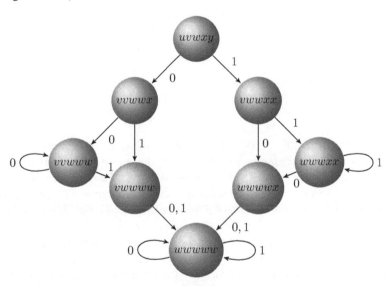

Figure 3.9. *Automaton of the C-IFS representing the subdivision of a third-degree NURBS curve. As with second-degree curves, every transition δ(uvw, 0) is associated with the subdivision matrix T_0(u, v, w, x, y) (see equation [3.11]) and every transition δ(uvwxy, 0) is associated with the matrix T_1(u, v, w, x, y) (see equation [3.12])*

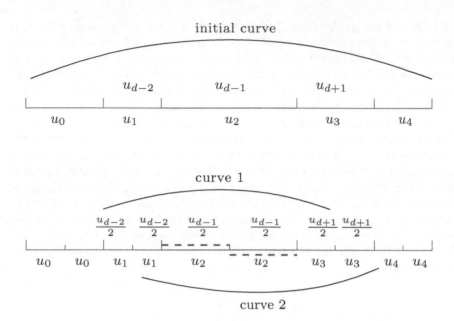

Figure 3.10. *The top diagram represents the knot interval vector of a curve of degree 3 for an initial curve*

COMMENTS ON FIGURE 3.10.– *The central interval, represented in blue, corresponds to the parameter values for which the curve is defined. The other intervals ($d-1$ intervals on each side) are useful in calculating the basis functions. After duplicating the knots, each interval is split (bottom diagram). Under the axis of the parameters, the lengths of the interval have been multiplied by 2, which does not change the definition of basis functions. The intervals on which the two curves are defined are identified with blue dotted lines. The knot interval vectors associated with each of these curves are obtained by completing each of the previous intervals by $d-1$ intervals on each side. The intervals $d-1$ at each end (represented in grey) are useless.*

We therefore deduce the algorithm for building the automaton (see Algorithm 3.1).

Now it merely remains to determine the values of the coefficients of the subdivision matrix associated with each transition. As we have seen for degrees 2 and 3, they are represented in the form of two types of matrices, parameterized by a knot interval vector, which can be instantiated for every case: $T_0^d(u_0, \cdots, u_{2d-2})$ and $T_1^d(u_0, \cdots, u_{2d-2})$. These matrices can be determined based on the formalism of the blossoms by calculating the expression of the new control points according to the old ones. The blossoms have been introduced by Ramshaw (1989) and can be seen as a generalization of the De Casteljau algorithm. We present the basis principle and the properties necessary to understand the computations of the subdivision matrix coefficients. For a NURBS curve $C(t)$, of degree d, defined by $d - 1$ control points and a knot vector $(t_0, \cdots, t_{2d-1})$, the principle is to associate it with a function with d variables $g(v_0, v_1, \cdots, v_{d-1})$ called a *polar form* verifying the following properties:

$-\ d$ is d-affine : $g(v_0, \cdots, t, \cdots, v_{d-1}) = \frac{(b-t)}{(b-a)} g(v_0, \cdots, a, \cdots, v_{d-1}) + \frac{(t-a)}{(b-a)} g(v_0, \cdots, b, \cdots, v_{d-1})$;

$-\ g$ is symmetrical: $g(v_0, \cdots, v_i, \cdots, v_j, \cdots, v_{d-1}) = g(v_0, \cdots, v_j, \cdots, v_i, \cdots, v_{d-1})$;

$-\ g$ verifies: $g(t, t, \cdots, t) = C(t)$.

Require: d ▷ Degree
Require: $U_{init} = (u_0, \cdots, U_{2d-2})$ ▷ knot interval vector
  1:   $StateList \leftarrow U_{init}$
  2:   DOUBLE(U_{init})
  3:   **procedure** DOUBLE(U)
  4:      $U_{Old} \leftarrow U$
  5:      $U \leftarrow$ Duplicate the values of U
  6:      $U_0 \leftarrow$ recover the $2d - 1$ values of U centered at the $(d - 1)$th value
  7:      **if** AddToAutomaton(U_0,U_{Old}) **then**
  8:          DUPLICATE(U_0)
  9:      **end if**
10:      $U_1 \leftarrow$ recover the $2d - 1$ values of U centered at the (d)th value
11:      **if** AddToAutomaton(U_1,U_{Old}) **then**
12:          DUPLICATE(U_1)
13:      **end if**
14:   **end procedure**
15:   **function** ADDTOAUTOMATON(U,U_{Origin})
16:      $Addition \leftarrow False$
17:      **if** $U \notin StateList$ **then**
18:          Add the state U to $StateList$
19:          $Add \leftarrow True$
20:      **end if**
21:      create a transition between the state U_{Origin} and state U
22:      **return** Add
23:   **end function**

Algorithm 3.1. *Algorithm for constructing the automaton for a NURBS curve of degree d*

For a polynomial (and even rational) curve, this form exists and is unique. This polar form has the remarkable property known as "consecutivity": $\forall i, g(t_i, t_{i+1}, \cdots, t_{i+d}) = P_i$.

EXAMPLE.– For a Bezier curve of degree 2:

$$C(t) = (1-t)^2 P_0 + 2t(1-t)P_1 + t^2 P_2$$

Its polar shape is given as:

$$\begin{aligned} g(u,v) = \; & (1-u)(1-v)P_0 \\ & + (u+v)(1 - \tfrac{u+v}{2})P_1 \\ & uv P_2 \end{aligned}$$

We can observe that g is affine in u and in v and that $g(u,v) = g(v,u)$. For a Bezier curve, the knot vector is $(0,0,1,1)$, and the control points then correspond to $g(0,0), g(0,1)$ and $g(1,1)$.

Only from these properties and without knowing the expression of the polar form, it is then possible to compute a point of the curve $C(t)$ from the control points $P_i = g(t_i, t_{i+1}, \cdots, t_{i+d})$.

The process consists of rewriting $g(t, \cdots, t)$ based on values of t_i in order to reconstruct sequences of d consecutive values of parameter t_i (that is, the control points P_i).

Consider the example of quadratic NURBS curves for a value of $t \in [t_i, t_i + 1[$:

$$g(t,t) = \frac{t_{i+1} - t}{t_{i+1} - t_i} g(t, t_i) + \frac{t - t_i}{t_{i+1} - t_i} g(t, t_{i+1})$$

which after rewriting in standard form, using the symmetry property, gives:

$$g(t,t) = \frac{t_i - t}{t_{i+1} - t_i} g(t_i, t) + \frac{t - t_{i+1}}{t_{i+1} - t_i} g(t, t_{i+1})$$

Then we continue, for the second value of t:

$$g(t, t_{i+1}) = \frac{t_{i+2} - t}{t_{i+2} - t_i} g(t_i, t_{i+1}) + \frac{t - t_i}{t_{i+2} - t_i} g(t_{i+2}, t_{i+1}) \tag{3.13}$$

$$= \frac{t_{i+2} - t}{t_{i+2} - t_i} g(t_i, t_{i+1}) + \frac{t - t_i}{t_{i+2} - t_i} g(t_{i+1}, t_{i+2}) \tag{3.14}$$

$$= \frac{t_{i+2} - t}{t_{i+2} - t_i} P_i + \frac{t - t_i}{t_{i+2} - t_i} P_{i+1} \tag{3.15}$$

$$g(t_i, t) = \frac{t_{i+1} - t}{t_{i+1} - t_{i-1}} g(t_i, t_{i-1}) + \frac{t - t_{i-1}}{t_{i+1} - t_{i-1}} g(t_i, t_{i+1}) \qquad [3.16]$$

$$= \frac{t_{i+1} - t}{t_{i+1} - t_{i-1}} g(t_{i-1}, t_i) + \frac{t - t_{i-1}}{t_{i+1} - t_{i-1}} g(t_i, t_{i+1}) \qquad [3.17]$$

$$= \frac{t_{i+1} - t}{t_{i+1} - t_{i-1}} P_{i-1} + \frac{t - t_{i-1}}{t_{i+1} - t_{i-1}} P_i \qquad [3.18]$$

The knot-doubling algorithm inserts new knots in the middle of each interval $[t_i, t_{i+1}[$. The new control points associated with this new knot vector are expressed using the polar form in the form $g(t_i, t_i^m)$ and $g(t_i^m, t_{i+1})$, where $t_i^m = \frac{t_i + t_{i+1}}{2}$.

By multi-affinity, the polar form allows the expression of a point of the curve $g(t, t)$ according to the set of control points $g(t_i, t_{i+1})$, but also according to any other set of reference points $g(x_i, x_{i+1})$. To represent the entire curve, certain constraints must be verified on the new set of knot interval values $\{x_i\}$. Once these reference points are chosen by means of the knot vector, using the multi-affinity property, we can express them according to the control points $g(t_i, t_{i+1})$. These reference points are the new control points associated with the new knot vector defining the curve $C(t)$.

In our case, the new knot vector is $(t_0^m, t_1, t_1^m, t_2, t_2^m)$, we do not have to worry about these constraints, the representativeness of the curve is guaranteed by the knot-doubling algorithm. If we reconsider equations [3.15] and [3.18], replacing t with t_i^m and utilizing the knot interval values $u_i = t_{i+1} - t_i$, we get:

$$g(t_i^m, t_{i+1}) = \frac{u_i + 2u_{i+1}}{2(u_i + u_{i+1})} P_i + \frac{u_i}{2(u_i + u_{i+1})} P_{i+1} \qquad [3.19]$$

$$g(t_i, t_i^m) = \frac{u_i}{2(u_i + u_{i-1})} P_{i-1} + \frac{u_i + 2u_{i-1}}{2(u_i + u_{i-1})} P_i \qquad [3.20]$$

And from there, we deduce the two matrices $T_0(uvw)$ and $T_1(uvw)$ given by expressions [3.9] and [3.10].

From the example of the second-degree curves, the generalization to any degree is quite simple. As a first step, we must determine the knot vectors for each curve, after doubling the knots. Figure 3.11 illustrates the approach. The central knot interval of the initial curve corresponds to the set of values of the parameter t for which the curve is defined. After knot-doubling, the initial curve is divided into two curves. Every knot interval is also subdivided into two. The central knot interval is split into two central knot intervals (seen in magenta dotted lines and green in the figure), one for each set of parameters of each new curve. To build the two complete knot vectors, it is sufficient to add the values of $d - 1$ knot intervals located on each side to each of these central knot values. We notice that on each side, there are $d - 1$ knot intervals that are not necessary to define the subdivided curves. In addition, depending on the parity of the degree, the first knot is either an old knot if the degree is odd or an inserted knot if the degree is even.

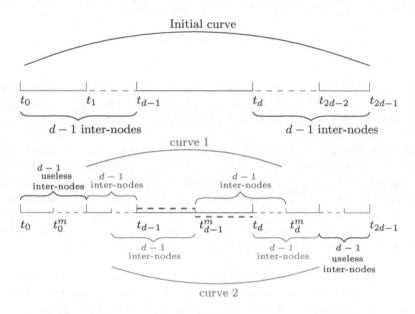

Figure 3.11. *Illustration of the extraction of the knot vectors from the two curves resulting from the knot-doubling of a curve of degree d*

COMMENTS ON FIGURE 3.11.– *At the top, the initial knot vector (in blue in the center) has a knot interval representing the values of the parameter t of the curve. At the bottom, knot-doubling subdivides the initial curve into two curves. Each knot interval is split into two intervals of the same size. The central knot interval is divided into two knot intervals. The first (represented as a magenta dotted line) corresponds to the parameter values of the first curve, and the second (represented as a green dotted line) to those of the parameter of the second curve. From each of these knot intervals, the d − 1 knot intervals of each side simply have to be considered to constitute the knot interval vector associated with each of the curves.*

Once the knot vectors are determined, we must calculate the new control points of the curve 1, based on the old ones, in order to obtain $T_0(u_0, u_1, \cdots, u_{d-2})$ and similarly, for the curve 2, in order to obtain $T_1(u_0, u_1, \cdots, u_{d-2})$. These new control points are identified by the different polar values that we can obtain by selecting d consecutive values of the new knot vectors. Therefore, we can obtain the different types of d-tuples of the following knot values:

– if d is odd:

$$(t_i, t_i^m, t_{i+1}, \cdots, t_{i+\frac{d-1}{2}-1}^m, t_{i+\frac{d-1}{2}}) \qquad [3.21]$$

$$(t_i^m, t_{i+1}, \cdots, t_{i+\frac{d-1}{2}}^m, t_{i+\frac{d-1}{2}}^m) \qquad [3.22]$$

– if d is even:

$$(t_i^m, t_{i+1}, \cdots, t_{i+\frac{d}{2}-1}^m, t_{i+\frac{d}{2}}) \qquad [3.23]$$

$$(t_i, t_i^m, \cdots, t_{i+\frac{d}{2}-1}, t_{i+\frac{d}{2}-1}^m) \qquad [3.24]$$

We thus have four possible configurations (two per parity type). Once these polar values are identified, we can calculate the expression of the new control points according to the old ones, using the multi-affine property of the polar form. The number of computational steps solely depends on the number of new knots (t_j^m) contained in the polar value vector: $\frac{d-1}{2}$ for the vector [3.21], $\frac{d+1}{2}$ for the vector [3.22] and $\frac{d}{2}$ for the vectors [3.23] and [3.24].

If the degree is high, the formal expression of the relation between the new control points and the old ones can be complex. However, from the knowledge of the different combinations of knot interval vectors produced by the algorithm for generating the automaton 3.2, it is easy to calculate each subdivision matrix for every state and a given initial knot interval vector. To this end, we need to build the "blossom" from each vector of polar values representing a new control point, by expressing it using the vectors of polar values corresponding to the old control points. The algorithm is based on a computational step consisting of "eliminating" an intermediate knot value t_k^m. Consider a vector of polar value in the form $(t_i, t_i^m, t_{i+1}, t_k^m, t_k^m, t_{k+1}, \cdots, t_j)$, where $i < k < j$ with i the smallest index, j the largest and k the index, such that every polar value greater than t_k^m are old knots. The decomposition consists of expressing the value of the polar form for this vector based on two others for which the value t_k^m is substituted by t_{i-1} for the first and by t_{i+1} for the second. Using the affinity and symmetry property to replace the polar values in the ascending order, we get the following relation:

$$
\begin{aligned}
P_k^m = {} & g(t_i, t_i^m, t_{i+1}, \cdots, t_k, \mathbf{t_k^m}, t_{k+1}, \cdots, t_j) \\
& - \frac{(t_{j+1} - t_k^m)}{(t_{j+1} - t_{i-1})} g(\mathbf{t_{i-1}}, t_i, t_i^m, t_{i+1}, \cdots, t_k, t_{k+1}, \cdots, t_j) \\
& + \frac{(t_k^m - t_j)}{(t_{j+1} - t_{i-1})} g(t_i, t_i^m, t_{i+1}, \cdots, t_k, t_{k+1}, \cdots, t_j, \mathbf{t_{j+1}})
\end{aligned}
$$

This expression allows one to build a tree (the blossom) per iteration, successively eliminating new knots, and deducing the expression of a new control point according to the old ones.

3.4.3. *Surface subdivision*

From the description of the NURBS curves by a C-IFS, it is easy to shift to surfaces. The subdivision transformations are calculated using the tensor product of the curve subdivision matrices. Regarding the automaton, we build it by achieving the equivalent of what we can call an automaton "tensor product" (see section 4.2.1 for a full presentation). It should be remembered that the surfaces are defined by the tensor product of two NURBS curves, usually of the same degree. Specifically, the tensor product is used to build the basis functions of the surfaces, from the basis functions of the curves. We therefore have a grid of control points $(P_{i,j}, w_{i,j})$ and two knot vectors $(s_0, \cdots, s_{m_s-1})$ and $(t_0, \cdots, t_{m_t-1})$. The NURBS surface is then defined by:

$$C(s,t) = \frac{\sum\limits_{i,j=0}^{m_s-1,m_t-1} P_{i,j} w_{i,j} N_{i,d}(s) N_{j,d}(t)}{\sum\limits_{i=0}^{m_s-1,m_t-1} w_{i,j} N_{i,d}(s) N_{j,d}(t)}$$

where $N_{i,d}(s)$ and $N_{j,d}(t)$ refer to the basis functions of each curve defined by the knot vectors $(s_0, \cdots, s_{m_s-1})$ and $(t_0, \cdots, t_{m_t-1})$.

The construction of the automaton is based on the combination of the two curve automatons. Figure 3.12 shows some of the combinations that we obtain during the subdivision process. The initial state consists of crossing the initial states $uvw \otimes u'v'w'$. For every curve, there are two transitions from the original state: one from the state uvw to the state uvv and one from the state uvw to the state vvw (respectively, from the state $u'v'w'$ to the state $u'v'v'$ and one from the state $u'v'w'$ to the state $v'v'w'$). These transitions translate the subdivision of each curve into two sub-curves. The combination of these states translates the subdivision of the surface into four subsurfaces. Thereby, let us define four transitions with the initial state $uvw \otimes u'v'w$ and as arrival states, the states $uvv \otimes v'v'w'$, $vvw \otimes u'v'v'$ and $vvw \otimes u'v'w'$, respectively. Each of these transitions is associated with a subdivision matrix obtained by making the tensor product of the two subdivision matrices associated with the corresponding curve transitions. Therefore, one simply has to follow the "unfolding" of the automatons of the two curves starting from the initial states in parallel. For every couple of transitions found, we create a state (named using the names of the destination states of the curves) and create a transition between the starting state and this new state. Finally, this transition is associated with a subdivision matrix, obtained by carrying out the tensor product of the subdivision matrices associated with the curve transitions. We recursively proceed with the construction by restarting from the pair of the two arrival states of the curves, until all of the states of the two curves have been traveled. Algorithm 3.2 is the translation of this construction. Figure 3.13 shows the automaton resulting from a bi-quadratic NURBS surface.

Require: Auto1, Auto2   ▷ two curve automatons
    $E_1Init \leftarrow$ initial state of $Auto1$
2: $E_2Init \leftarrow$ initial state of $Auto2$
    E Create state $E_1Init \otimes E_2Init$
4: CROSSSTATE(E_1Init, E_2Init, E)
    procedure CROSSSTATE(E_1, E_2, E_{State})
6:    **for** for every transition T_1 of E_1 **do**
        for for every transition T_2 of E_2 **do**
8:            $E_1^D \leftarrow$ recover the destination state of T_1
            $E_2^D \leftarrow$ recover the destination state of T_1
10:        **if** $E_1^D \otimes E_2^D \notin AutoRes$ **then**
            create and add $E_1^D \otimes E_1^D$ to $AutoRes$
12:            create the transition $T_1 \otimes T_2$ between E_{Start} and $E_1^D \otimes E_1^D$
            Associate with $T_1 \otimes T_2$ matrix calculated by tensoriel product
between the matrix T_1 and matrix T_2
14:        **end if**
            CROSSSTATE($E_1^D, E_2^D, E_1^D \otimes E_2^D$)
16:        **end for**
    end for
18: **end procedure**

Algorithm 3.2. *Algorithm for constructing the automaton for a NURBS surface from the automatons of the two NURBS curves*

3.5. Subdivision curves and surfaces

Subdivision curves and surfaces are constructed using an iterative calculation process. We can then expect that there is a link with the C-IFS and BC-IFS.

Subdivision surfaces are built starting from a surface defined by a mesh. The latter is described by a B-rep-type topological structure (reflecting the mesh connectivity, that is, the relations between faces, edges and vertices). An iterative computation process will allow this mesh to be gradually "refined" by constructing a sequence of meshes $\{M_i\}_{i \in \mathbb{N}}$ such that $\lim_{i \rightarrow \infty} M_i = smooth\,surface$. The initial mesh is referred to as a control mesh (similarly to NURBS). It is considered to be a crude representation of an underlying boundary area computed by an iterative refinement process. As iterations progress, the mesh becomes denser by the addition of vertices, edges and faces. The positions of the new vertices are precisely determined to converge toward a smooth surface. Most of the time, these computations are defined to converge to a chosen boundary surface, such as a Bezier surface and a B-spline surface. To achieve this refinement, we need to define two types of processing:

– build the topology of mesh M_{i+1}, based on that of M_i;

– compute the positions of the vertices of the new mesh, based on the vertices of the old mesh.

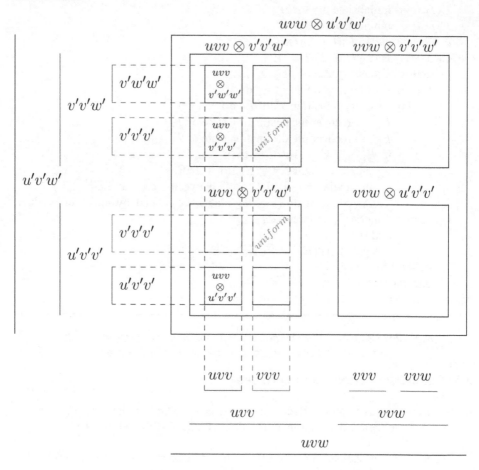

Figure 3.12. *Subdivision process of a NURBS surface of degree 2, obtained by the tensor product of two NURBS curves of degree 2*

The rules of topological construction are usually described by explaining how to add vertices, edges and faces based on the previous structure. For example, for each face and edge, a new vertex must be inserted. Then, we have to specify how the B-rep structure is built from these vertices: for each face, we must create an edge between the new face point and the new points of the edges bordering the face, etc. The computation rules are provided based on "masks" representing the neighborhood around the new point, enhanced with a weighting (determining the calculation of the new point from the old ones).

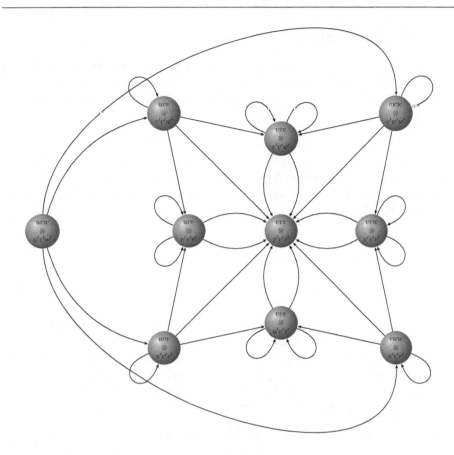

Figure 3.13. *Automaton of the C-IFS representing the subdivision of a NURBS surface of degree 2, defined by the tensor product of two curves of degree 2. The automaton was built using Algorithm 3.2*

We will see that we can represent subdivision surfaces using BC-IFS. The representation of subdivision surfaces using BC-IFS helps to explain known properties, but generally not integrated into conventional implementations. On one hand, the BC-IFS encode the topological subdivision of surfaces and, on the other hand, they introduce an additional element: the evaluation primitive. We will see that it is not just a question of the "purity" of the model and that these concepts allow one to solve real problems, such as the construction of junctions between the subdivision surfaces of different natures.

3.5.1. *Subdivision curves*

For curves, the control "meshing" is reduced to a polygon. It follows that the problem is much simpler because of the trivial B-rep structure of a polygon: a vertex is incident to two edges for the internal vertices, and to a single edge for the boundaries.

3.5.1.1. *Example of uniform quadratic B-spline curves*

Generally for curves, topological structuring and the computation of the positions of the new points are reduced to the data of the computational rules of the new points from the old ones. The indices implicitly define the topological structure of the new polygon (a list of points is sufficient):

$$P_{2i}^{n+1} = \frac{3}{4}P_i^n + \frac{1}{4}P_{i+1}^n \qquad [3.25]$$

$$P_{2i+1}^{n+1} = \frac{1}{4}P_i^n + \frac{3}{4}P_{i+1}^n \qquad [3.26]$$

This rule is implicitly tantamount to inserting two vertices at the level of every segment, removing the old vertices P_{i+1}^n and if $i > 0$, transforming them into edges (joining P_{2i+1}^{n+1} to P_{2i+2}^{n+1}). In practice, we simply calculate the list of the new vertices. However, for surfaces, it is no longer possible to proceed implicitly. These rules are only applicable if we have the minimum number of control points (which is $d + 1$, where d represents the degree of the curve $d + 1 = 3$).

The matrix formulation of this calculation is given as:

$$(P_0^{n+1} P_1^{n+1} P_2^{n+1} P_3^{n+1}) = (P_0^n P_1^n P_2^n) \begin{pmatrix} 3/4 & 1/4 & 0 & 0 \\ 1/4 & 3/4 & 3/4 & 1/4 \\ 0 & 0 & 1/4 & 3/4 \end{pmatrix} \qquad [3.27]$$

REMARK.– Classically, equation [3.27] is presented in the transposed form; in other words, with the right multiplication by a column vector of control points (the matrix being transposed). We present it in this form so that it is in agreement with the presentation of the C-IFS.

From there, we can easily deduce the subdivision process:

$$(P_0^{n+1} P_1^{n+1} P_2^{n+1}) = (P_0^n P_1^n P_2^n) \begin{pmatrix} 3/4 & 1/4 & 0 \\ 1/4 & 3/4 & 3/4 \\ 0 & 0 & 1/4 \end{pmatrix} \qquad [3.28]$$

$$(P_1^{n+1} P_2^{n+1} P_3^{n+1}) = (P_0^n P_1^n P_2^n) \begin{pmatrix} 1/4 & 0 & 0 \\ 3/4 & 3/4 & 1/4 \\ 0 & 1/4 & 3/4 \end{pmatrix} \qquad [3.29]$$

This decomposition brings forward the subdivision process of uniform quadratic B-spline curves and we naturally find the Chaikin matrices.

3.5.1.2. *Example of uniform cubic B-spline curves*

For uniform cubic B-spline curves, the principle is the same. Subdivision rules are written as:

$$P_{2i}^{n+1} = \frac{1}{8}P_{i-1}^n + \frac{3}{4}P_i^n + \frac{1}{8}P_{i+1}^n \qquad [3.30]$$

$$P_{2i+1}^{n+1} = \frac{1}{2}P_i^n + \frac{3}{2}P_{i+1}^n \qquad [3.31]$$

To build a cubic curve, we must have at least four control points. By applying the subdivision rules to these four initial points, we get five points for the refined polygon. In matrix form, this calculation assumes the following form:

$$(P_0^{n+1} P_1^{n+1} P_2^{n+1} P_3^{n+1} P_4^{n+1}) = (P_0^n P_1^n P_2^n P_3^n) \begin{pmatrix} \frac{1}{2} & \frac{1}{8} & 0 & 0 & 0 \\ \frac{1}{2} & \frac{3}{4} & \frac{1}{2} & \frac{1}{8} & 0 \\ 0 & \frac{1}{8} & \frac{1}{2} & \frac{3}{4} & \frac{1}{2} \\ 0 & 0 & 0 & \frac{1}{8} & \frac{1}{2} \end{pmatrix} \qquad [3.32]$$

This expression can be broken down by applying two operators:

$$(P_0^{n+1} P_1^{n+1} P_2^{n+1} P_3^{n+1}) = (P_0^n P_1^n P_2^n P_3^n) \begin{pmatrix} \frac{1}{2} & \frac{1}{8} & 0 & 0 \\ \frac{1}{2} & \frac{3}{4} & \frac{1}{2} & \frac{1}{4} \\ 0 & \frac{1}{8} & \frac{1}{2} & \frac{3}{4} \\ 0 & 0 & 0 & \frac{1}{8} \end{pmatrix}$$

$$(P_1^{n+1} P_2^{n+1} P_3^{n+1} P_4^{n+1}) = (P_0^n P_1^n P_2^n P_3^n) \begin{pmatrix} \frac{1}{8} & 0 & 0 & 0 \\ \frac{3}{4} & \frac{1}{2} & \frac{1}{8} & 0 \\ \frac{1}{8} & \frac{1}{2} & \frac{3}{4} & \frac{1}{2} \\ 0 & 0 & \frac{1}{8} & \frac{1}{2} \end{pmatrix} \qquad [3.33]$$

3.5.1.3. *Uniform B-spline curves: subdivision* versus *BC-IFS*

The approach, previously illustrated for quadratic and cubic B-spline curves, is easily generalized to any degree d.

We have a set of rules (in finite number) that determine the calculation of a vector P^{n+1} of $d + 2$ control points, using a vector P^n of $d + 1$ control points. By decomposing P^{n+1} into two vectors P_G^{n+1} and P_D^{n+1} (respectively, composed of the $d + 1$ first points and the $d + 1$ last ones), using the rules we can express P_G^{n+1} and P_D^{n+1} in the form: $P_G^{n+1} = T_G P_n$ and $P_D^{n+1} = T_D P_n$. The attractor of the IFS

$\{T_G, T_D\}$ projected according to vector of control points P^0, corresponds to the subdivision curve that is represented by the C-IFS automaton, as shown in Figure 3.14.

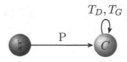

Figure 3.14. *C-IFS automaton whose attractor is a subdivision curve built from the polygon $P = (P_0, P_1, \cdots, P_{n-1})$. The matrices associated with the transitions T_D and T_G are derived from the subdivision matrix conventionally named by S in the context of subdivision surfaces*

We find the curve structures defined by BC-IFS in the structure of matrices T_G and T_D, with a sub-block of dimension $d \times d$ in the top left of T_G and the same sub-block in the bottom right of T_D. This sub-block corresponds to the subdivision operator of the vertices at the ends of the curve. The first vertex depends on the first d control points and the second on the last d control points.

If we analyze the calculations made when we recursively apply the subdivision rules to obtain successive refinements, we find that they correspond to those of the evaluation tree of the corresponding C-IFS (see Figure 3.14), but without the application of the evaluation primitive K. For l refinement steps, we should have calculated all of the possible combinations $PT_{\sigma_0}T_{\sigma_1} \cdots T_{\sigma_{l-1}}$ for $\sigma_i \in \{D, G\}$. The difference is that with the rule-based formulation, there is no redundant computation, unlike C-IFS due to the common columns in the matrices T_D and T_G. Nonetheless, this disadvantage disappears and turns into the advantage of the C-IFS model, when we introduce the evaluation primitive K. Indeed, one of the problems with subdivision curves (as with surfaces) is that for a small number of refinements, the refined polygon approximating the boundary curve may be significantly distant from the latter. Especially when the vertices of the approximating polygon do not belong to the boundary curve. By choosing, as primitive K, the line segment connecting the fixed points of each transformation T_D and T_G, the union of the transformations of k by $PT_{\sigma_0}T_{\sigma_1} \cdots T_{\sigma_{l-1}}$ are segments whose vertices belong to the curve (regardless of the value of $l \geq 0$). In addition, this time, the redundancy of the calculations is limited to the vertices shared by each segment. These properties can be exploited to provide generic and efficient evaluation algorithms for implementation on Graphics Processing Units (GPUs) (Lawlor 2012; Morlet *et al.* 2018).

3.5.2. *Subdivision surfaces*

The main interest of subdivision surfaces is to allow the construction of smooth surfaces from meshes of control points whose structure is not regular. As we mentioned in the introduction to this section, the refinement of the initial mesh is based on two types of rules: to define how the topology of the initial mesh needs to be modified in order to obtain the topology of the refined mesh, and to explain the calculation of the coordinates of the vertices of the new mesh, according to the old one. Generally, the rules for building the new topology are independent of the irregularity of the mesh. On the other hand, the computational rules depend on this irregularity, in particular to ensure an order of continuity in the neighborhood of irregular vertices. We will present two subdivision schemes: the Doo–Sabin and the Catmull–Clark schemes. The first generalizes the regular case of uniform biquadratic B-spline surfaces and the second, bicubic uniform B-spline surfaces.

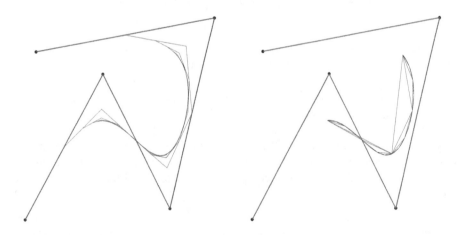

Figure 3.15. *Approximations of a uniform cubic B-spline curve for levels 0 to 6: on the left, using subdivision rules, on the right by BC-IFS*

COMMENTS ON FIGURE 3.15.– *The B-spline curve is defined by five control points. At level 0, the approximation using the subdivision rules is equal to the starting polygon, and using BC-IFS , the approximation connects the vertices of the pieces of the boundary curves (totaling two for this example). For the first approximation levels and due to the primitive K, the BC-IFS model produces approximations "closer" to the boundary curve than the subdivision rule computation.*

3.5.2.1. *Doo–Sabin scheme*

The Doo–Sabin scheme was originally designed to extend uniform quadratic B-spline surfaces in order to control meshes of any topology. The regular case then corresponds to uniform quadratic B-spline surfaces and is applied to a mesh of regular control points, possessing at least $3 \times 3 = 9$ control points (defining quadrangular faces). The refinement produces a mesh of $4 \times 4 = 16$ control points, as illustrated by Figure 3.16.

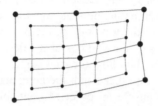

Figure 3.16. *Refinement of a regular control mesh. In red: Minimal control mesh for the regular case of the Doo–Sabin scheme. In green: First refinement level (subdivision points)*

To understand and highlight the iterative process, one should note that the refinement computation will proceed in the same way at each refinement, applying the same computation rules for each new subgroup of 3×3 control points obtained at the previous stage. Therefore, starting from the 4×4 control points obtained after the first refinement, we can identify four sub-meshes of 3×3 new control points, as illustrated by Figure 3.17. Each of the sub-meshes corresponds to the control points of a quarter of the limit surface. We then use the Chaikin algorithm, which was presented in section 3.2 for curves, but is now applied to surfaces. The automaton representing this subdivision is then composed of a state S and four transitions (see Figure 3.18).

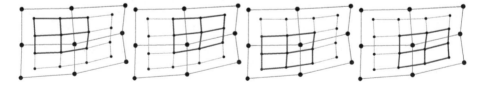

Figure 3.17. *Self-similarity of the refined mesh. Each of the four blue sub-meshes represents the control points of a quarter of the limit surface*

The subdivision matrices, T_0^s, T_1^s, T_2^s and T_3^s, associated with each subdivision are then simply obtained from the subdivision rules expressing the new control points from the old ones. They can also be obtained from the tensor products of the subdivision matrices T_0^c (see equation [3.28]) and T_1^c (see equation [3.29]) of curves: $T_0^s = T_0^c \otimes T_0^c$; $T_1^s = T_0^c \otimes T_1^c$; $T_2^s = T_1^c \otimes T_0^c$ and $T_3^s = T_1^c \otimes T_1^c$.

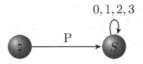

Figure 3.18. *Automaton of a subdivision surface for the Doo–Sabin scheme*

REMARK.– This result could have been obtained by applying the construction properties per tensor product of two uniform quadratic B-spline subdivision curves.

For the Doo–Sabin scheme, irregular cases appear as soon as one of the vertices of the control mesh is not of valence 4. The refinement process will then generate an irregular face (where the number of edges is equal to the valence of the irregular vertex). However, every new vertex will be of valence 4. After this first refinement, the irregular faces remain irregular, and keep the same number of edges. No other irregular face is created. Assume that we have applied a first refinement. Let us take the example of an irregular face with three edges (a triangle). As shown in Figure 3.19, a new refinement reveals a single triangle and quadrangular faces. We can then identify the decomposition process of the control mesh: an irregular mesh comprising the triangle and three other regular meshes with 3×3 control points (see Figure 3.20). Each of these meshes represents a quarter of the boundary area. During a new refinement stage, the regular mesh will follow the subdivision process of the regular case, and the irregular mesh will be subdivided into an irregular mesh and three regular meshes. The automaton for the irregular case is presented by Figure 3.21.

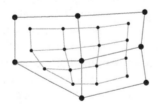

Figure 3.19. *Refinement of an irregular control mesh. In red: Minimal control mesh for the regular case of the Doo–Sabin scheme. In green: First refinement level (subdivision points)*

In order to implement subdivision surfaces using iterated function systems, there is no need to introduce incidence and adjacency constraints. The topological structure is implicit to the subdivision process. Subdivision matrices deduced from the subdivision rules incorporate these constraints and guarantee the topology. Only one C-IFS is enough. It remains to choose a primitive K to calculate an approximation of the surface. In the regular case, as we have seen for curves, a relevant choice is to take a quadrangular face whose four vertices are the fixed points of the four subdivision matrices T_0^R, T_1^R, T_2^R and T_3^R. For the irregular case, a vertex will be replaced by the fixed point of the subdivision matrix T_0^I, associated with the transition looping on the irregular state.

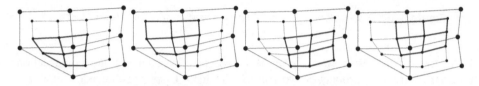

Figure 3.20. *Self-similarity of the refined mesh. In blue, the four sub-meshes of which one is irregular, representing the control points of four quarters of the boundary surface*

Figure 3.21. *Automaton of a surface subdivision for the Doo–Sabin scheme with an irregular face*

As we have just mentioned, the explicit coding of the topology is not necessary. Nonetheless, it is interesting to explain it to build connections between two subdivision surfaces or between a subdivision surface and any fractal surface. In the uniform case, a surface tile defined by nine control points is bordered by four curves, each defined by six control points. We find the incidence and adjacency constraints between the subdivided tiles, as shown in Figure 3.22. These constraints defining the topology are the same as a quadrangular surface tile subdividing into four subsurfaces. The topological structure and topological subdivision structure are identical to those presented in section 2.3.1. The irregular tile is bordered by two irregular curves and two regular curves. Irregular curves are themselves bordered by a regular vertex on one side and an irregular vertex on the other side. The incidence and adjacency constraints for these curves are almost identical to those of regular curves except that the two types of vertices (regular and irregular) are defined by two distinct states. The same is true for constraints on surfaces with one state for each type of curves. The complete BC-IFS automaton performing the topological subdivision of an irregular quadrangular tile is given by Figure 3.25.

The cellular decomposition associated with the subdivision process of an irregular tile is presented in Figure 3.26. We can note that it is very similar to that of a regular tile (see Figure 1.52). However, the cells involved in the scheme are not always of the same type. In order to have a complete description, it is necessary to add that of the cell decomposition of regular tiles R (see Figure 1.52) with $a = b = CR$.

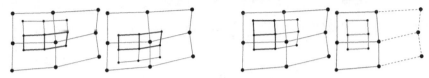

Figure 3.22. *Adjacency and incidence constraints. On the left: Example of an adjacency constraint between the right-hand side sub-tiles. The two subsets of subdivision points share common points (mesh in thick-line): the bottom edge of the top tile must be equal to the top edge of the bottom tile. On the right: Example of an incidence constraint. The left edge of the top tile must be equal to the subdivision (top) of the left curve*

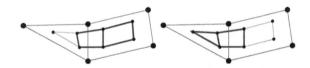

Figure 3.23. *Control points defining one of the irregular edges. The irregular edges are bordered by an irregular vertex on one of the sides and a regular vertex on the other side. The left-hand side vertex is defined by three control points, connected in green in the left diagram. The regular vertex is defined by four control points, connected in green in the right diagram*

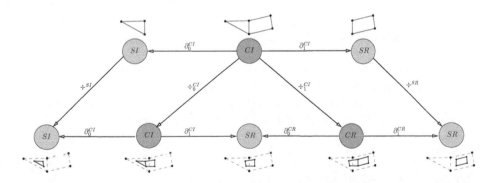

Figure 3.24. *Illustration of incidence and adjacency constraints for an irregular curve CI with an irregular vertex SI on the left and a regular vertex SR on the right. The irregular curve is subdivided into an irregular curve on the left and a regular curve CR on the right*

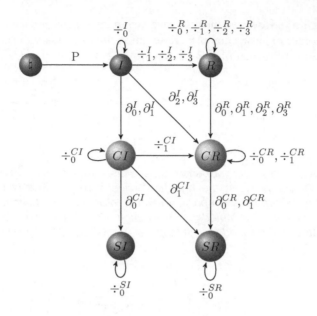

Figure 3.25. *Representation of the topological subdivision process of an irregular tile (with an irregular vertex) by the automaton of a BC-IFS*

REMARK.– For the sake of clarity, we chose to only use one type of irregular curve I and one type of regular curve, but we could have chosen the more general case of two types of irregular curves and two types of regular curves, analogously to the cellular decomposition addressed for the regular case shown in Figure 1.52.

From this cell decomposition, we can explain the incidence and adjacency constraints (see Appendix, section A.2).

For a tile with an irregular vertex of a different nature (for example, originating from a mesh refinement with a vertex of valence other than three), the topological structure, as well as the constraints, are strictly identical. Only the dimensions of the spaces associated with the states of the irregular vertices, curves and faces and thus the sizes of the subdivision matrices change. The coefficient values are obviously different.

3.5.2.2. *Catmull–Clark scheme*

The Catmull–Clark scheme generalizes uniform cubic B-spline surfaces of meshes with any topology. At the first refinement level, we obtain a mesh with quadrangular faces. Irregularity cases appear for vertices of valence other than four. The formalization of this scheme using the BC-IFS is rigorously identical to that of the Doo–Sabin scheme. For the regular case, the Catmull–Clark scheme requires a mesh of 4×4 control points. A refinement step transforms this mesh into a mesh of 5×5 control points (see Figure 3.28). From this refinement, we can extract four sub-meshes of 4×4 control points, from which we can reapply the process (see Figure 3.29).

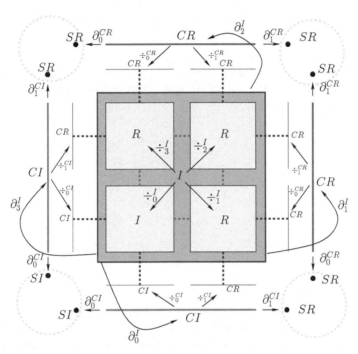

Figure 3.26. *Cell decomposition of an irregular tile. The constraints are represented by dotted lines: blue for incidences, red for adjacencies and adjacencies on incidence operators*

The automaton representing this subdivision is then the same as that of the Doo–Sabin scheme (see Figure 3.18).

The irregular case arises when one of the four central points of the mesh of 4×4 control points has a valence other than four. Figure 3.30 illustrates the subdivision of the irregular mesh into three regular meshes and one irregular mesh. Therefore, the automaton describing this decomposition is once again identical to that of the irregular case of the Doo–Sabin scheme. Following the same reasoning as that conducted for the Doo–Sabin scheme, we can show that the cell decomposition is also the same (see Figure 3.26). The same incidence and adjacency constraints are then derived (see Appendix, section A.2). The complete automaton is shown in Figure 3.25. Only the dimensions of the spaces associated with the states differ.

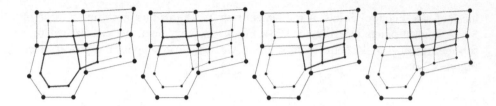

Figure 3.27. *Subdivision of an irregular tile with an irregular face of six sides. The subdivision process is identical to the irregular case with a triangular face: the irregular tile is subdivided into an irregular tile and three regular tiles. The incidence and adjacency constraints remain the same*

Figure 3.28. *Catmull subdivision of a regular tile*

3.5.2.3. *Loop scheme*

The Loop scheme is based on the subdivision of a triangular mesh. Each triangle of the control mesh is divided into four, by inserting a vertex for each edge (see Figure 3.31(a)). The topology of the new mesh is thus composed of the old and the new vertices. For each new vertex, its position is calculated according to the vertices of the edge from which it originates and the two vertices opposite to that edge (see Figure 3.31(b)). The position of the old vertices is recalculated according to the vertices of its 1-neighborhood of the old mesh (see Figure 3.31(c)). The regular vertices are those with a valence equal to 6. Thereby, the minimal configuration to build a piece of triangular elemental surface consists of 12 control points (see Figure 3.32).

The topological subdivision is the standard subdivision of a triangle with edges of the same type in the regular case (see Figure 2.23). Figure 3.33 shows the subdivision points for the four subdivided faces. The C-IFS is therefore composed of a "Face" state, from which four transitions originate and lead to the same state. The transformations associated with these transitions transform the initial red mesh into each of the refined meshes represented in blue. By choosing an arbitrary numbering for the control points (see Figure 3.32), and using the computation rules presented by Figure 3.31, we can then easily determine the four subdivision transformations. As an example, we give the transformation matrix T_0 (see Figure 3.33(a)):

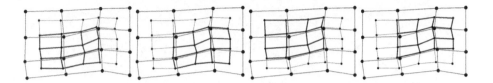

Figure 3.29. *The four sub-meshes of* 4×4 *control points are represented in blue from which we can reapply the process*

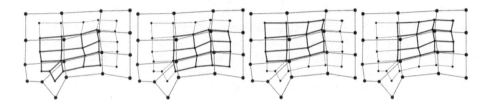

Figure 3.30. *Example of an irregular mesh with a vertex of valence five. After refinement, the mesh is decomposed into an irregular mesh of the same type and three other regular meshes*

$$T_0 = \begin{pmatrix} \frac{3}{8} & \frac{1}{8} & 0 & \frac{1}{8} & \frac{1}{16} & 0 & 0 & 0 & 0 & 0 & 0 \\ \frac{1}{8} & \frac{3}{8} & \frac{3}{8} & 0 & \frac{1}{16} & \frac{1}{8} & \frac{1}{16} & 0 & 0 & 0 & 0 \\ 0 & 0 & \frac{1}{8} & 0 & 0 & 0 & \frac{1}{16} & 0 & 0 & 0 & 0 \\ \frac{1}{8} & 0 & 0 & \frac{3}{8} & \frac{1}{16} & 0 & 0 & \frac{1}{8} & 0 & \frac{1}{8} & 0 & 0 \\ \frac{1}{8} & \frac{3}{8} & \frac{1}{8} & \frac{3}{8} & \frac{10}{16} & \frac{3}{8} & \frac{1}{16} & \frac{3}{8} & \frac{3}{8} & 0 & \frac{1}{8} & \frac{1}{16} \\ 0 & \frac{1}{8} & \frac{3}{8} & 0 & \frac{1}{16} & \frac{3}{8} & \frac{10}{16} & 0 & \frac{1}{8} & \frac{3}{8} & 0 & \frac{1}{16} \\ 0 & 0 & 0 & 0 & 0 & 0 & \frac{1}{16} & 0 & 0 & 0 & 0 & 0 \\ 0 & 0 & 0 & \frac{1}{8} & \frac{1}{16} & 0 & 0 & \frac{3}{8} & \frac{1}{8} & 0 & \frac{3}{8} & \frac{1}{16} \\ 0 & 0 & 0 & 0 & \frac{1}{16} & \frac{1}{8} & \frac{1}{16} & \frac{1}{8} & \frac{3}{8} & \frac{3}{8} & \frac{3}{8} & \frac{10}{16} \\ 0 & 0 & 0 & 0 & 0 & 0 & \frac{1}{16} & 0 & 0 & \frac{1}{8} & 0 & \frac{1}{16} \\ 0 & 0 & 0 & 0 & 0 & 0 & 0 & 0 & 0 & 0 & \frac{1}{8} & \frac{1}{16} \\ 0 & 0 & 0 & 0 & 0 & 0 & 0 & 0 & 0 & 0 & 0 & \frac{1}{16} \end{pmatrix}$$ [3.34]

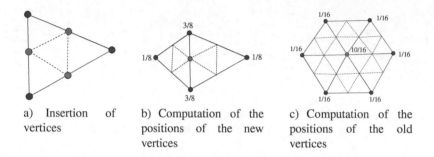

a) Insertion of vertices

b) Computation of the positions of the new vertices

c) Computation of the positions of the old vertices

Figure 3.31. *Illustration of the Loop subdivision. Figure 3.31(a) is an illustration of the topological refinement of the mesh: A new point (in red) is inserted for each edge of the old mesh and the old points are preserved. As a result, four triangular faces are created from one face of the initial mesh. Figures 3.31(b) and 3.31(c) present the computation rules for the positions of the vertices of the refined mesh*

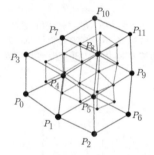

Figure 3.32. *Refinement of a regular control mesh for the Loop scheme. The structure of the mesh of the minimal control points used to build an elementary patch using the Loop diagram is represented in red. The control points have been arbitrarily numbered. This numbering allows the subdivision matrix given by equation [3.35] to be written. The mesh after a subdivision stage is shown in green*

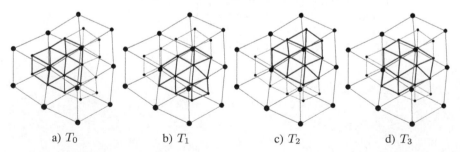

a) T_0 b) T_1 c) T_2 d) T_3

Figure 3.33. *Self-similarity of the refined mesh for the Loop scheme. Decomposition of the mesh subdivided (in green) into four meshes (in blue), with the same configuration as the original mesh (in red). The subdivision can be reapplied independently to each of these four new meshes. This decomposition highlights the property of self-similarity of the surface.*

4

Building Operations, Assistance to Design and Applications

As we have seen in the examples of Chapter 2, using the BC-IFS model, we can relatively simply describe the topology of self-similar curves or wired structures. For surfaces, the expression of incidence and adjacency constraints is slightly more tedious and requires more commitment. For volume subdivisions, the complexity increases. It may become more challenging to formulate the topological definition of a BC-IFS, without incidence and adjacency constraints, leading to inconsistencies. These inconsistencies result in a system of equations without a solution.

It is possible, however, to integrate conditions of symmetry into the model that facilitate this design stage and make it possible to remove some of these constraints.

A complementary solution, to simplify the designer's task, is to propose a graphical interface. By symbolizing the cellular structure, it is then possible to interactively define the incidence and adjacency constraints (see Figure 4.1). By adding a set of conventions on the cell decomposition, according to an identified bias, the description of the topology can be simplified.

It is also possible to propose construction operators, capable of combining basic structures, to automatically obtain the definition of a more sophisticated structure. These operators are largely widespread and widely used in CAD. As examples, we can mention extrusion or Boolean operations (intersection, union, difference, etc.).

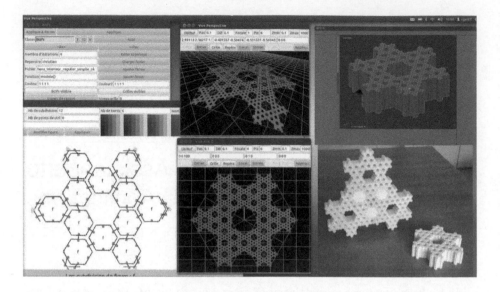

Figure 4.1. *Example of the user interface of "MODITERE", the iterative modeler developed jointly by the LIB and the LIRIS (source: project MODITERE no. ANR-09-COSI-014). For a color version of this figure, see www.iste.co.uk/gentil/geometric.zip*

COMMENTS ON FIGURE 4.1.– *At the bottom right, the incidence and adjacency relations can be described interactively, based on a symbolic representation of the cellular decomposition. In the center, the overall geometry can be controlled by means of control points (in red), and the local geometry (the definition of subdivision matrices) is directly controlled from subdivision points (in blue). Once the topology and geometry are defined, the model can be exported, either to be reused in a modeler (Blender here) or manufactured directly.*

4.1. Topological consistency and symmetry constraints

4.1.1. *Orientation constraints*

As we first mentioned, at the end of section 2.3, incidence and adjacency constraints must satisfy a specific consistency, otherwise the topological structure may be degenerated (for example, reduced to a single point). These consistencies are of different types.

The first type corresponds to adjacency constraints on incidence operators (see section 1.3.5). These are of the same kind as the cellular decomposition constraints of conventional B-rep structures. For example, a face must be bordered by edges, themselves bordered by vertices, and two successive edges have to share the same vertex. Models such as simplicial complexes ensure this consistency. The BC-IFS model is more general and the principle is to describe the parts of the structures we wish to control, without necessarily describing everything. The consistency of the structure is then the responsibility of the designer.

The second type of consistency is related to the orientation of cells when we define adjacency or incidence constraints. Let us consider the example of the definition of a connection between two subdivisions of a face through their edges. The adjacency relation of the type $\div_1^F \partial_2^F = \div_2^F \partial_0^F$ (see Figure 4.2) pools the edges (edge 0 with edge 2) with the whole associated cellular structure, that is the vertex 0 of the edge 0 with the vertex 0 of the edge 2 and vertex 1 of the edge 0 with vertex 1 of the edge 2. These constraints also appear for incidence relations. It is then necessary to build a consistent system of constraints, even if it means generating degenerate situations, as shown in Figure 4.2.

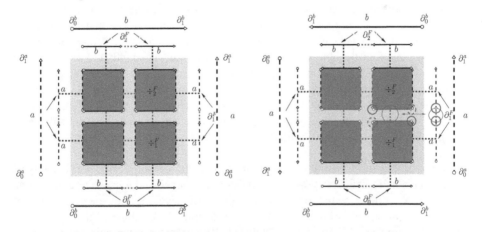

Figure 4.2. *Example of orientation constraints of edges for the definition of the quadrangular subdivision of a solid face. For a color version of this figure, see www.iste.co.uk/gentil/geometric.zip*

COMMENTS ON FIGURE 4.2.– *The arrows symbolize the orientation of the edges with a circle at the origin of the arrow indicating the edge 0 (∂_0^a and ∂_0^b); and at the opposite end, an arrow indicating the edge 1 (∂_1^a and ∂_1^b). On the left, the orientation of the edges chosen at the outset does not generate contradictions when establishing the incidence and adjacency constraints. While on the right, we note an inversion of the central edges (identified in red). The adjacency constraint itself is not a problem. It is as if the subdivided face was "twisted" to connect to the other face. However, this "twist" will generate a degeneration because of the incidence relation of the edge of* *the "right-hand border", which constrains the "twisted" face to stick to the "right-hand border" of the face: the vertices indicated by a circle in solid green line must be identical; those surrounded by a dotted green line must also be identical due to the adjacency relation, but they must also be shared with the previous vertices by the incidence relation.*

Let us take the example of a quadrangular subdivision of a face (illustrated on the left in Figure 4.2). In this figure, the orientations of the edges a and b have been indicated by a circle and an arrow, respectively, symbolizing the border (or vertex) 0 and 1 of the edge. We have chosen a direction for which two opposing edges have their circles and their arrows, respectively, positioned on the same side (on the right, respectively, on the left for b, and at the top, respectively, at the bottom for a). Therefore, the subdivided faces will have edges with orientations compatible with both incidence and adjacency relations. For the configuration on the right-hand side of Figure 4.2, the orientation was chosen by turning the face around. The edges a and b have been placed alternately, always starting with the circle and ending with the arrow. Therefore, during the subdivision process, we can observe that the edges of the subdivided elements have the same orientation as the subdivided edges. Incidence relations, identified by blue dotted lines, then generate no conflict. For adjacency relations (indicated by red dotted lines), we note that the orientations of the edges are opposite. It is not a problem in itself. This constraint will result in a "twist" of the face. However, this same subdivided face must also follow the edge a of the face. As a result, incidence and adjacency relations, indicated by the red circles, will be in conflict. There are also other conflicts with the other adjacency constraints.

The designer must then find a configuration of the orientations of the face edges, so as to be able to define a system of consistent incidence and adjacency constraints. For example, to describe the topology of Figure 1.26, a solution for the orientation of the edges is presented in Figure 4.3. Consistency must be established throughout the subdivision process, here for the two states of the automaton referencing one another.

4.1.2. *Permutation operators and symmetry constraints*

Guaranteeing the consistency of orientations can be facilitated by the introduction of permutation operators and symmetry constraints *vis-à-vis* these permutations. For this purpose, we proceed in three steps:

1) the definition of the permutation operator;

2) the definition of the symmetry constraint, making the cell invariant by permutation;

3) the use of the operator in an adjacency or incidence constraint.

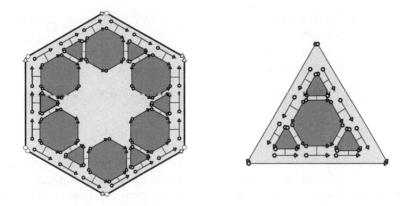

Figure 4.3. *Example of configuration of edge orientation for the attractor of Figure 1.26, representing the intersection of a Menger cube with a plane. Incidence relations are indicated in blue and adjacency relations in red. For a color version of this figure, see www.iste.co.uk/gentil/geometric.zip*

4.1.2.1. *Edges*

Consider, for example, the definition of an adjacency relation between two faces via an edge. In order for these faces to connect to one another, regardless of the orientation of their edges (see Figure 4.4), we need to define a permutation operator.

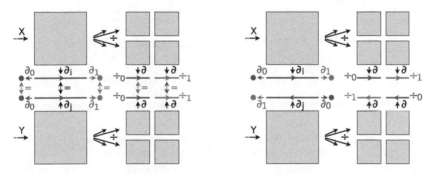

Figure 4.4. *Issues in the orientation of edges. On the left, the connection for which the edges are oriented in the same direction. On the right, a configuration in which the edges involved in the connection are of different orientations. In the latter case, a conventional adjacency constraint $X\partial_i = Y\partial_j$ will result in a twist of the face. For a color version of this figure, see www.iste.co.uk/gentil/geometric.zip*

The permutation operator is represented by a permutation matrix specifying how spaces are interchanged. We define it using equations describing its action, as shown in the left-hand diagram of Figure 4.5. After applying the permutation on the edge (at the bottom), the border ∂_1 (green vertex) must match the border ∂_0 (red vertex of the edge without permutation, at the top). Conversely, the border ∂_0 of the edge with permutation must match the border ∂_1 of the edge without permutation. These constraints are expressed by the following two equations:

$$P\partial_0 = \partial_1 \qquad\qquad\qquad [4.1]$$

$$P\partial_1 = \partial_0 \qquad\qquad\qquad [4.2]$$

where P represents the permutation matrix. If the edge has a non-zero internal dimension, it is also necessary to specify their mapping after application of the permutation. To this end, we use internal operators, represented by a column vector I_i. These internal operators identify the dimension associated with the ith control point of the cell that is not part of any border. In this case, since there is only one internal dimension, after application of the permutation it is sent onto itself. Therefrom, we deduce the additional equation:

$$PI_0 = I_0 \qquad\qquad\qquad [4.3]$$

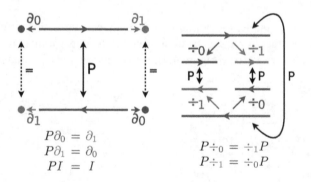

Figure 4.5. *Definition and application of the permutation operator. On the left, the definition of the permutation (a symmetry, here) on a curve. On the right, application of the symmetry constraint on the curve. For a color version of this figure, see www.iste.co.uk/gentil/geometric.zip*

Boundary and internal operators are composed of only 0 and 1, with only one 1 in each column. The resolution of equations [4.1], [4.2] and [4.3], defining the permutation, is then immediate. For our example, $\partial_0 = (1,0,0)^\top$, $\partial_1 = (0,0,1)^\top$ and $I_0 = (0,1,0)^\top$. We deduce that:

$$P = \begin{pmatrix} 0 & 0 & 1 \\ 0 & 1 & 0 \\ 1 & 0 & 0 \end{pmatrix}$$

Once the permutation P is defined, we have to write the symmetry constraints to make the edge invariant by P. We then refer to the central diagram of Figure 4.5, illustrating the effect of applying the permutation during the subdivision process. After application of the permutation, the subdivisions end up inverted, in "position" (right-/left-hand side in the figure) and in orientation. Therefore, the permutation ought to be used again to make the orientations compatible. We deduce the following equations: $P\div_0 = \div_1 P$ and $P\div_1 = \div_0 P$.

Let A denote the attractor of the IFS $= \{\div_0, \div_1\}$ associated with the edge:

$$
\begin{aligned}
P\mathcal{A} &= P(\div_0\mathcal{A} \cup \div_1\mathcal{A}) \\
&= P(\div_0\mathcal{A}) \cup P(\div_1\mathcal{A}) \\
&= P(P(\div_1\mathcal{A})) \cup P(P(\div_0\mathcal{A})) \\
&= P(P(\div_1\mathcal{A} \cup \div_0\mathcal{A}) \\
&= P(P(A))
\end{aligned}
$$

If P is an involution, that is, $PP = Id$, then $P\mathcal{A} = \mathcal{A}$; in other words, $\mathcal{A}$ is invariant by P. The edge can then be used indiscriminately to achieve direct connections (without changing the orientation, if the configuration allows it) or by introducing the permutation P if necessary. Figure 4.6 shows an example of connection via opposite orientation edges. Here, the adjacency relations connect the border of a subdivided face with the permuted border of another subdivided face. The equations are $X\partial_i = Y\partial_j P$, where X and Y can represent any other symbol, either a subdivision or a border symbol. The equation then represents the following situation:

– an adjacency relation $\div_x\partial_i = \div_y\partial_j P$ (inverted connection between two subdivided faces as in the example of Figure 4.2 with $X = \div_1^F$ and $Y = \div_2^F$);

– an adjacency relation on the incidence operators $\partial_x\partial_i = \partial_y\partial_j P$ (a volume bordered by two faces whose connection on an edge is inverted).

The resolution of the equations of the permutation constraints yields the following constraints on the coefficients of the subdivision operators.

By denoting: $\div_0 = \begin{pmatrix} a_0 & b_0 & c_0 \\ a_1 & b_1 & c_1 \\ a_2 & b_2 & c_2 \end{pmatrix} \quad \div_1 = \begin{pmatrix} d_0 & e_0 & f_0 \\ d_1 & e_1 & f_1 \\ d_2 & e_2 & f_2 \end{pmatrix}.$

By applying the equations of the permutation constraints, we have:

$$
P\div_0 = \begin{pmatrix} 0 & 0 & 1 \\ 0 & 1 & 0 \\ 1 & 0 & 0 \end{pmatrix} \begin{pmatrix} a_0 & b_0 & c_0 \\ a_1 & b_1 & c_1 \\ a_2 & b_2 & c_2 \end{pmatrix} = \begin{pmatrix} a_2 & b_2 & c_2 \\ a_1 & b_1 & c_1 \\ a_0 & b_0 & c_0 \end{pmatrix}
$$

$$\div_1 P = \begin{pmatrix} d_0 & e_0 & f_0 \\ d_1 & e_1 & f_1 \\ d_2 & e_2 & f_2 \end{pmatrix} \begin{pmatrix} 0 & 0 & 1 \\ 0 & 1 & 0 \\ 1 & 0 & 0 \end{pmatrix} = \begin{pmatrix} f_0 & e_0 & d_0 \\ f_1 & e_1 & d_1 \\ f_2 & e_2 & d_2 \end{pmatrix}$$

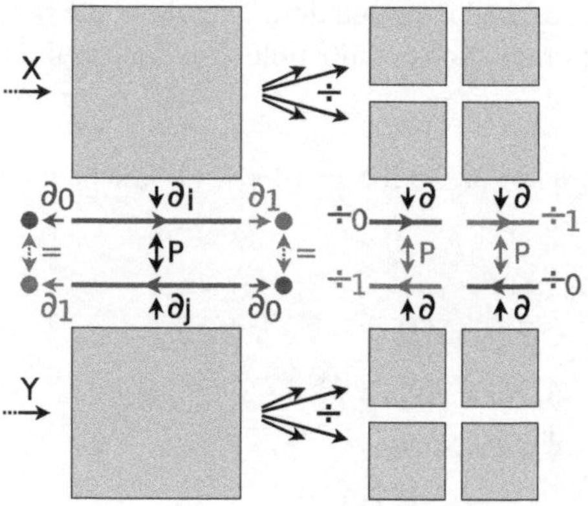

$$X\partial_i = Y\partial_j P$$

Figure 4.6. *Construction of a connection between two edges of opposite orientations. For a color version of this figure and the preceding equations, see www.iste.co.uk/gentil/geometric.zip*

COMMENTS ON FIGURE 4.6.– *On the left, the adjacency relation between the two faces is set up using the permutation P defined in Figure 4.5. On the right, we find that the invariance of the edge by P, imposed by the permutation constraints on the edge (Figure 4.5 on the right side), ensures the consistency of the subdivision system. The adjacency constraint equation $X\partial_i = Y\partial_j P$ is defined after a subdivision step by:* $X\partial_i \div_0 = Y\partial_j \div_1 P$ *and* $X\partial_i \div_1 = Y\partial_j \div_0 P$.

We obtain two subdivision matrices presenting a "central symmetry" of their coefficients (for a color version of these equations, see the link above):

$$\div_0 = \begin{pmatrix} a_0 & b_0 & c_0 \\ a_1 & b_1 & c_1 \\ a_2 & b_2 & c_2 \end{pmatrix} \qquad \div_1 = \begin{pmatrix} c_2 & b_2 & a_2 \\ c_1 & b_1 & a_1 \\ c_0 & b_0 & a_0 \end{pmatrix}$$

REMARK.– As with the example presented, the quadratic Bezier curves are defined from three control points. Their vertices depend on a single control point and they have an internal dimension equal to 1. They are symmetrical and are a special case of the example being addressed. As a result, the De Casteljau matrices have the following symmetrical structure:

$$D_0 = \begin{pmatrix} 1 & \frac{1}{2} & \frac{1}{4} \\ 0 & \frac{1}{2} & \frac{1}{2} \\ 0 & 0 & \frac{1}{4} \end{pmatrix} \qquad D_1 = \begin{pmatrix} \frac{1}{4} & 0 & 0 \\ \frac{1}{2} & \frac{1}{2} & 0 \\ \frac{1}{4} & \frac{1}{2} & 1 \end{pmatrix}$$

4.1.2.2. *Vertices*

The question of the orientation that we have illustrated with the edges is similarly expressed with the other cells, such as the faces or other structures with fractal topology. It is also present at the vertex level. A vertex may depend on several control points. The influence of these control points on the vertex is not necessarily symmetrical, hence a notion of orientation. Internal operators can somehow be used to index the internal dimensions of a cell by associating an index (the operator index) with a dimension of the barycentric space. Let us take the example of a curve a that is divided into two parts, depending on five control points, whose vertices each depend on two control points. For the curve, we define an internal operator $I_0^a = (0,0,1,0,0)$. For vertices, which are necessarily defined by the same state and subdivision operator, we define two internal operators $I_0^s = (1,0)^\top$ and $I_1^s = (0,1)^\top$.

The symmetry operator of the vertex is defined by equations $P^s I_0 = I_1$ and $P^s I_1 = I_0$. We deduce thereof that $P^s = \begin{pmatrix} 0 & 1 \\ 1 & 0 \end{pmatrix}$. The symmetry constraint on the vertex is then: $P^s \div^s = \div^s P^s$. This equation yields that $\div^s = \begin{pmatrix} a & 1-a \\ 1-a & a \end{pmatrix}$.

We can make use of the fact that the vertex is symmetrical and that two vertices of this type can be connected in any "direction" that is with or without application of symmetry. Figure 4.7 shows, on the left, an example of a direct connection without using symmetry (the adjacency constraint is $\div_0^a \partial_1^a = \div_1^a \partial_0^a$); in the center and on the right side, the figure shows an example of a connection using the permutation (the adjacency constraint is $\div_0^a \partial_1^a = \div_1^a \partial_0^a P^s$). We can observe that the inversion of one of the two vertices in the connection will induce a cusp point at the center of the curve (while in the case of the direct connection, we have continuity of the derivative). This cusp point is repeated by self-similarity. To illustrate this property, we have been careful in selecting subdivision matrices in order that the tangents to the curves at the end points are carried by the segments defined by the control points of the vertices.

4.1.2.3. *Faces and volumes*

For planar structures, it is not always easy to find a consistent orientation of the edges that satisfies the adjacency and incidence relations. For volume structures connecting each other by means of faces, the problem is even more complex. The use of permutation operators becomes almost indispensable. Figure 4.8 shows examples of configurations that we may encounter when defining a volume subdivision system. We may note that the faces F that we want to connect are not oriented in the same way. In the configuration of the left diagram, if we connect the faces directly, without a permutation operator, we shall obtain cusp areas. For the center and right configurations, we shall obtain a twist of the volume cell. Moreover, if these density cells are otherwise constrained and these constraints are not compatible, we shall obtain a degeneration of the cell.

We may note that we can resolve all three cases if we define a permutation that swaps two vertices and leaves the third unaffected, and a circular permutation of the vertices in either direction (see Figure 4.7). From these two permutations and by composition we are able to address all of the situations.

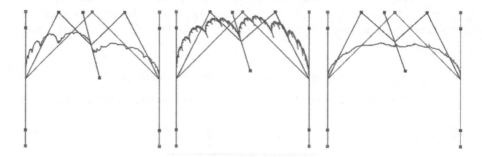

Figure 4.7. *The three curves have a single internal dimension, and vertices of dimension 2. For a color version of this figure, see www.iste.co.uk/gentil/geometric.zip*

COMMENTS ON FIGURE 4.7.– *For the left-hand curve, the adjacency relation has been conventionally defined without introducing a permutation. For the center curve and the one on the right, we have applied a permutation to one of the vertices to inverse it before connecting it:* $\div_0^a \partial_1^a = \div_1^a \partial_0^a P^s$. *We can observe that at the connection level, the curve turns backwards, thus defining a cusp point. This cusp point is duplicated all over the curve by self-similarity.*

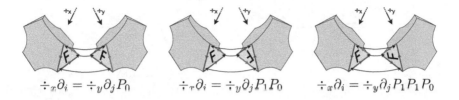

$$\div_\pi \partial_i = \div_y \partial_j P_0 \qquad \div_\pi \partial_i = \div_y \partial_j P_1 P_0 \qquad \div_x \partial_i = \div_y \partial_j P_1 P_1 P_0$$

Figure 4.8. *The definition of the connections for volume cells can prove complex. The use of permutations facilitates the designer's work when the faces to be connected do not have the same orientation. For a color version of this figure, see www.iste.co.uk/gentil/geometric.zip*

The definition of the first permutation P_0^F is given by:

$$P_0^F \partial_0 = \partial_0 P^a$$
$$P_0^F \partial_1 = \partial_2 P^a$$
$$P_0^F \partial_2 = \partial_1 P^a$$
$$P_0^F I_0 = I_0$$

The operator P_0^F swaps the two bottom vertices on the left figure and leaves the top vertex invariant. P^a represents the permutation making the edge symmetrical and is not described in these equations; for its definition, refer to section 4.1.2.2. We assume the face has only one internal dimension identified by I_0. We note on the top left-hand diagram in Figure 4.9 that if we overlay the face and the permuted face, the directions of the edges do not coincide. To keep the orientations consistent, after applying the permutation of the face, we must apply the permutations of the edges. Hence equations of the form: $P_0^F \partial_x = \partial_y P^a$.

Still in Figure 4.9, the diagram on the bottom left shows a subdivision step. It enables us to deduce the permutation constraints that must be satisfied by the face to be invariant by P_0^F. The equations are then:

$$P_0^F \div_0 = \div_0 P_0^F$$
$$P_0^F \div_1 = \div_2 P_0^F$$
$$P_0^F \div_2 = \div_1 P_0^F$$
$$P_0^F \div_3 = \div_3 P_0^F$$

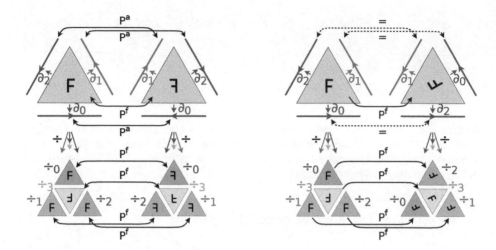

Figure 4.9. *Illustration of the definitions of face permutations: on the left, a vertical axis symmetry; on the right, a rotation of one-third of a turn to the left. For a color version of this figure, see www.iste.co.uk/gentil/geometric.zip*

The second permutation P_1^F is defined and can be similarly applied to the first of the right-hand diagrams. The equations for defining P_1^F are as follows:

$$P_1^F \partial_0 = \partial_1$$
$$P_1^F \partial_1 = \partial_2$$
$$P_1^F \partial_2 = \partial_0$$
$$P_1^F I = I$$
$$P_1^F \div_0 = \div_1 P_1^F$$
$$P_1^F \div_1 = \div_2 P_1^F$$
$$P_1^F \div_2 = \div_0 P_1^F$$
$$P_1^F \div_3 = \div_3 P_1^F$$

The volume elements shown in Figure 4.8 can then be connected by the following adjacency relations using the permutations:

$$\div_x \partial_i = \div_y \partial_j P_0$$
$$\div_x \partial_i = \div_y \partial_j P_1 P_0$$
$$\div_x \partial_i = \div_y \partial_j P_1 P_1 P_0$$

4.2. Topological combination

As we have seen from the design examples of Chapter 2, notation of incidence and adjacency relations can be tedious. For example, the definition of a quadrangular surface that is subdivided into 9 tiles requires 12 incidence equations, 12 adjacency equations and 4 adjacency equations on incidence operators. Although the writing appears to be systematic, it requires attention and is time-consuming. Therefore, in some cases, we can define topological combination operators that automate this writing process, simplifying the designer's approach. These are indeed topological operators since they will operate on the definition of the automaton, as well as incidence and adjacency equations.

In section 3.4.3, we have already mentioned the generation of surfaces from the tensor products of curves (see Algorithm 3.2). We have proposed a way to automatically define the automaton, building the surface based on the automatons of the two curves. However, the incidence and adjacency constraints have not been explained. They were induced by the structures of the surface subdivision matrices, and calculated using the tensor products of the curve subdivision matrices. In this case, the result corresponds to a surface defined by tensor product.

In this section, we propose to generalize this approach by defining, on the one hand, the tensor product of BC-IFS, taking into account both automatons and incidence and adjacency constraints. Here, subdivision matrices will not be set and the result of the operation is the definition of a topology.

We present a second operator to build tree structures such as those presented in section 2.5. The input data are an arbitrary BC-IFS, which represents the leaves of the tree structure that we want to build. The operator generates the complement of the BC-IFS, defining the topology of the tree structure.

4.2.1. *Topological tensor product*

The topological tensor product must provide a topological representation of the classic geometric tensor product; in other words, it is obtained by multiplying the basic functions.

Therefore, from two BC-IFS provided on input, the topological tensor product operator generates a new BC-IFS for which:

– the associated automaton reflects the subdivision process of the geometric tensor product;

– incidence and adjacency relations reflect the topology of the geometric tensor product.

Figure 4.10 symbolizes two examples of a topological tensor product: on the left, a curve by a curve, and on the right, a Cantor set by a Sierpinski triangle. The geometric tensor product is a particular case of the topological tensor product that corresponds to the same topological structure, but for which subdivision matrices are obtained from the tensor products of the subdivision matrices of the initial BC-IFS. For the topological tensor product, incidence and adjacency relations will induce dependencies between subdivision matrices. The degrees of freedom will allow the geometry to change while maintaining the topology.

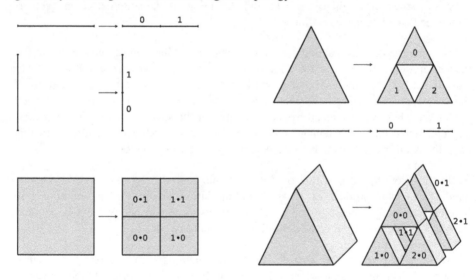

Figure 4.10. *Examples of construction of the topological tensor product. For a color version of this figure, see www.iste.co.uk/gentil/geometric.zip*

COMMENTS ON FIGURE 4.10.– *The diagrams on the left illustrate the topological tensor product of a curve by a curve, and the diagrams on the right that of a Cantor set by a Sierpinski triangle. In both cases, the subdivision processes of each BC-IFS provided as input are combined to build the tensor product.*

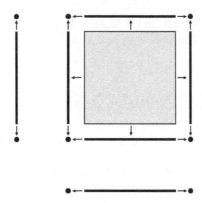

Figure 4.11. *Example of cellular decomposition of a tensor product of two curves revealing the incidences*

We denote by $(Q_1, \Sigma_1, \delta_1, \gamma_1) \bullet (Q_2, \Sigma_2, \delta_2, \gamma_2)$ the two BC-IFS provided on input. The BC-IFS resulting from the topological tensor product operation is denoted by $(Q, \Sigma, \delta, \gamma)$. The operator will be denoted as $\bullet$: $(Q, \Sigma, \delta, \gamma) = (Q_1, \Sigma_1, \delta_1, \gamma_1) \bullet (Q_2, \Sigma_2, \delta_2, \gamma_2)$.

We then need to define the set of states Q, the set of symbols Σ, the transition function δ and the equivalency relations γ, expressed from the BC-IFS on input.

REMARK.– To simplify the language, and by analogy to the geometric tensor product, we will talk about "directions". Traditionally, for a surface, these directions represent the parameters s and t of basic functions $B(s)$ and $B(t)$ of each curve. The basic function of the surface built by tensor product $B(s,t) = B(s) \otimes B(t)$ is defined, excepting two parameters, making it possible to move on the surface according to the directions linked to each of the parameters u and v.

4.2.1.1. *States*

Similar to the construction of the tensor product between two curves, the cells that compose each of the curves (curves and vertices) are combined to one another to form new cells: the curves are combined to form a quadrangular surface, the combination of the vertex of one of the curves with another curve forms a curve (vertex $\bullet$ curve) and that of a vertex with another vertex forms a vertex (vertex $\bullet$ vertex). Thus, each cell of the first BC-IFS is combined with all those of the second. In the case of automatons, these cells are symbolized by states. Thereby, the set of the states of Q is built as a Cartesian product of the two sets of states of the initial automatons. Each pair of states $(x,y) \in Q_1 \times Q_2$ is associated with a state of Q:

$$Q = \{x \bullet y \mid (x,y) \in Q_1 \times Q_2\} \qquad\qquad [4.4]$$

Here, we denote the operator in the same way, associating every pair of states of $Q_1 \times Q_2$ to a state of Q. The number of states of Q is $|Q| = |Q_1| \times |Q_2|$.

4.2.1.2. Symbols

The symbols are constructed in a similar manner to the states. We denote by $\bullet$ the operator associating the symbols of Σ_1 and Σ_2 to the symbols of Σ. The symbols either represent subdivisions or incidence operators. For subdivision symbols of a state $x \bullet y$ of Q, we must apply all combinations of subdivisions originating from x and y.

Therefore, each pair of subdivision symbols $(u, v) \in \Sigma_{\div}^x \times \Sigma_{\div}^y$ is associated with a subdivision symbol of $\Sigma_{\div}^{x \bullet y}$:

$$\forall (x, y) \in Q_1 \times Q_2 \, , \\ \Sigma_{\div}^{x \bullet y} = \{ \div_{u \bullet v}^{x \bullet y} \mid (\div_u^x, \div_v^y) \in \Sigma_{\div}^x \times \Sigma_{\div}^y \} \tag{4.5}$$

The number of subdivision symbols is then $|\Sigma_{\div}^{x \bullet y}| = |\Sigma_{\div}^x| \times |\Sigma_{\div}^y|$.

For incidence symbols, the approach is slightly different. For a state $x \bullet y$, we have two ways of identifying an incidence: either in one direction (namely following one of the two BC-IFS following the state x, for example) or following the other (following the state y). With each incidence symbol of $\Sigma_{\partial}^x \cup \Sigma_{\partial}^y$, we associate an incidence symbol of $\Sigma_{\partial}^{x \bullet y}$.

For this construction, we define a neutral symbol ε for each set Σ^x and Σ^y. These symbols will be used to be combined to the right and left side with the incidence symbols, proceeding as follows:

$$\forall (x, y) \in Q_1 \times Q_2 \\ \Sigma_{\partial}^{x \bullet y} = \{ \partial_{i \bullet \varepsilon}^{x \bullet y} \mid \partial_i^x \in \Sigma_{\partial}^x \} \cup \{ \partial_{\varepsilon \bullet j}^{x \bullet y} \mid \partial_j^y \in \Sigma_{\partial}^y \} \tag{4.6}$$

Finally, we add the symbol $\varepsilon^{x \bullet y}$ from the two neutral symbols of Σ^x and Σ^y:

$$\forall (x, y) \in Q_1 \times Q_2 \, ; \quad \varepsilon^{x \bullet y} = \varepsilon \bullet \varepsilon \tag{4.7}$$

The number of incidence symbols is then $|\Sigma_{\partial}^{x \bullet y}| = |\Sigma_{\partial}^x| + |\Sigma_{\partial}^y|$.

4.2.1.3. Transitions

The destination state of a combination of two subdivision symbols is naturally the combination of the destination states of these same symbols.

Following the transition $\div_{u \bullet v}^{x \bullet y}$ of Q is equivalent to simultaneously following the transitions $\div_u^x$ of Q_1 and $\div_v^y$ of Q_2. We thus have:

$$\forall (x, y) \in Q_1 \times Q_2 \, ; \, \forall (\div_u^x, \div_v^y) \in \Sigma_{\div}^x \times \Sigma_{\div}^y \\ \delta(x \bullet y \, , \, \div_{u \bullet v}^{x \bullet y}) = \delta_1(x \, , \, \div_u^x) \bullet \delta_2(y \, , \, \div_v^y) \tag{4.8}$$

However, for incidence symbols, as mentioned above in their definition, the incidence is defined only in one of the two directions, the other direction being identified by the neutral symbol ε. The destination state merely has to be determined by following the associated transition of Q_1 or Q_2, depending on the position of the neutral element in the incidence symbol ($i \bullet \varepsilon$ or $\varepsilon \bullet j$):

$$\forall (x,y) \in Q_1 \times Q_2 \; ; \; \forall \partial_i \in \Sigma_\partial^x$$
$$\delta(x \bullet y \, , \, \partial_{i \bullet \varepsilon}^{x \bullet y}) = \delta_1(x \, , \, \partial_i^x) \bullet y \qquad\qquad [4.9]$$

$$\forall (x,y) \in Q_1 \times Q_2 \; ; \; \forall \partial_j \in \Sigma_\partial^y$$
$$\delta(x \bullet y \, , \, \partial_{\varepsilon \bullet j}^{x \bullet y}) = x \bullet \delta_2(y \, , \, \partial_j^y) \qquad\qquad [4.10]$$

4.2.1.4. *Equivalence relations*

For equivalence relations, the principle is relatively simple. When an equivalence relation exists in one direction, it must be propagated in the other direction to all cells that have been constructed from the initial cells involved. The notation expressing this construction is a little more technical but is simplified by the introduction of a new operator $\circ$ acting between a symbol and a path.

We recursively define $\circ$ by composing the symbols of the left path to the right. Different cases are distinguished according to the types of symbols encountered (edge, subdivision or neutral).

These cases are, in order, subdivision-subdivision, subdivision-edge, edge-edge, neutral-edge and edge-empty path:

$$\forall(\div_i^x, \div_j^y, \theta) \in \Sigma_\div^x \times \Sigma_\div^y \times L$$
$$\div_i^x \circ (\div_j^y \, \theta) \; = \; (\div_{i \bullet j}^{x \bullet y}) \, (\varepsilon \circ \theta)$$

$$\forall(\div_i^x, \partial_j^y, \theta) \in \Sigma_\div^x \times \Sigma_\partial^y \times L$$
$$\div_i^x \circ (\partial_j^y \, \theta) \; = \; (\partial_{\varepsilon \bullet i}) \, (\div_i^x \circ \theta)$$

$$\forall(\partial_i^x, \partial_j^y, \theta) \in \Sigma_\partial^x \times \Sigma_\partial^y \times L$$
$$\partial_i^x \circ (\partial_j^y \, \theta) \; = \; (\partial_{i \bullet \varepsilon}) \, (\partial_{\varepsilon \bullet j}) \, (\varepsilon \circ \theta)$$

$$\forall(\partial_j, \theta) \in \Sigma_\partial^y \times L$$
$$\varepsilon \circ (\partial_j \, \theta) \; = \; (\partial_{\varepsilon \bullet j}) \, (\varepsilon \circ \theta)$$

$$\forall \partial_i^x \in \Sigma_\partial^x$$
$$\partial_i^x \circ \varepsilon \; = \; \partial_{i \bullet \varepsilon}$$

The operator $\circ$ is defined in a similar way between a path in a left argument and a symbol in a right argument.

For each state $x \bullet y$ of Q, the adjacency relations derived from that state are built by a combination of adjacency relations (if any) originating from each state x and y.

$$\forall (x,y) \in Q_1 \times Q_2$$
$$\gamma_{\div}^{x \bullet y} = \{(\theta_1 \circ \div_u^y , \theta_2 \circ \div_u^y) \mid$$
$$(\theta_1, \theta_2) \in \gamma_{\div}^x , \div_u^y \in \Sigma_{\div}^y \} \cup \qquad\qquad [4.11]$$
$$\{(\div_u^x \circ \theta_1 , \div_u^x \circ \theta_2) \mid$$
$$(\theta_1, \theta_2) \in \gamma_{\div}^y , \div_u^x \in \Sigma_{\div}^x \}$$

The number of adjacency relations is $|\gamma_{\div}^{x \bullet y}| = |\gamma_{\div}^x| \times |\Sigma_{\div}^y| + |\gamma_{\div}^y| \times |\Sigma_{\div}^x|$.

The adjacency relations on incidence operators derived from a state $x \bullet y$ of Q are built independently from the adjacency relations on the incidence operators originating from the two states x and y:

$$\forall (x,y) \in Q_1 \times Q_2$$
$$\gamma_{\partial}^{x \bullet y} = \{(\theta_1 \circ \varepsilon^y , \theta_2 \circ \varepsilon^y) \mid (\theta_1, \theta_2) \in \gamma_{\partial}^x \} \cup$$
$$\{(\varepsilon^x \circ \theta_1 , \varepsilon^x \circ \theta_2) \mid (\theta_1, \theta_2) \in \gamma_{\partial}^y \} \cup \qquad\qquad [4.12]$$
$$\{(\partial_{i^x \bullet \varepsilon^y} \partial_{\varepsilon^{\delta_1(x,i^x)} \bullet j^y} , \partial_{\varepsilon^x \bullet j^y} \partial_{i^x \bullet \varepsilon^{\delta_2(y,j^y)}}) \mid$$
$$(\partial_{i^x}, \partial_{j^y}) \in \Sigma_{\partial}^x \times \Sigma_{\partial}^y \}$$

The number of adjacency relations on incidence operators is $|\gamma_{\partial}^{x \bullet y}| = |\gamma_{\partial}^x| + |\gamma_{\partial}^y| + |\Sigma_{\partial}^x| \times |\Sigma_{\partial}^y|$.

4.2.2. *Example of surface generation from curves*

Consider the case of two curves a and b, divided into two parts. The automaton of the BC-IFS representing the curve a is presented in Figure 4.12.

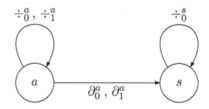

Figure 4.12. *Automaton of the curve a*

The formulation of the transitions of this automaton is then:

$$\delta(a, \div_0^a) = a$$
$$\delta(a, \partial_0^a) = s$$
$$\delta(a, \div_1^a) = a$$
$$\delta(a, \partial_1^a) = s$$
$$\delta(s, \div_0^s) = s$$

For curve a, the incidence and adjacency relations are given by:

$$\gamma^a = \gamma^a_{\div} = \left\{ \begin{array}{l} (\ \div^a_0 \ \partial^a_0 \ , \ \partial^a_0 \ \div^s_0\ ) \\ (\ \div^a_1 \ \partial^a_1 \ , \ \partial^a_1 \ \div^s_0\ ) \\ (\ \div^a_0 \ \partial^a_1 \ , \ \div^a_1 \ \partial^a_0\ ) \end{array} \right\}$$

The quotient graph is presented in Figure 4.13.

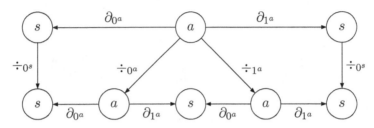

Figure 4.13. *Quotient graph of curve* a

The curve b is defined in the same way. In all equations, one simply substitutes the letter b for the letter a and the letter t (representing the state of the vertex of the curve b) for the letter s (representing the state of the vertex of the curve a).

4.2.2.1. *States and transitions*

Surface states are automatically generated: $Q = \{a \bullet b, a \bullet t, s \bullet b, s \bullet t\}$.

For each state, the transitions are constructed from equations [4.8]–[4.10]. The resulting automaton is illustrated by Figure 4.14.

$$\varepsilon^{a\bullet b} = \varepsilon^a \bullet \varepsilon^b \ ; \ \ \varepsilon^{a\bullet t} = \varepsilon^a \bullet \varepsilon^t \ ; \ \ \varepsilon^{s\bullet b} = \varepsilon^s \bullet \varepsilon^b \ ; \ \ \varepsilon^{s\bullet t} = \varepsilon^s \bullet \varepsilon^t$$

4.2.2.2. *The adjacency relation*

Adjacency relations $\gamma^{a\bullet b}_{\div}$ are built from the rules of [4.12]. The following relations are obtained when considering one direction, that is by combining $\gamma^a_{\div}$ with $\Sigma^b_{\div}$:

	$\div^b_0$
$(\ \div^a_0 \ \partial^a_0 \ , \ \partial^a_0 \ \div^a_0\ )$	$(\ \div^{a\bullet b}_{0\bullet 0} \ \partial^{a\bullet b}_{0\bullet\varepsilon} \ , \ \partial^{a\bullet b}_{0\bullet\varepsilon} \ \div^{s\bullet b}_{0\bullet 0}\ )$
$(\ \div^a_1 \ \partial^a_1 \ , \ \partial^a_1 \ \div^a_0\ )$	$(\ \div^{1^a\bullet 0^b}_{1^a\bullet 0^b} \ \partial^{a\bullet b}_{1\bullet\varepsilon} \ , \ \partial^{a\bullet b}_{1\bullet\varepsilon} \ \div^{s\bullet b}_{0\bullet 0}\ )$
$(\ \div^a_0 \ \partial^a_1 \ , \ \div^a_1 \ \partial^a_0\ )$	$(\ \div^{a\bullet b}_{0\bullet 0} \ \partial^{a\bullet b}_{1\bullet\varepsilon} \ , \ \div^{a\bullet b}_{1\bullet 0} \ \partial^{a\bullet b}_{0\bullet\varepsilon}\ )$

	$\div^b_1$
$(\ \div^a_0 \ \partial^a_0 \ , \ \partial^a_0 \ \div^a_0\ )$	$(\ \div^{a\bullet b}_{0\bullet 1} \ \partial^{a\bullet b}_{0\bullet\varepsilon} \ , \ \partial^{a\bullet b}_{0\bullet\varepsilon} \ \div^{s\bullet b}_{0\bullet 1}\ )$
$(\ \div^a_1 \ \partial^a_1 \ , \ \partial^a_1 \ \div^a_0\ )$	$(\ \div^{a\bullet b}_{1\bullet 1} \ \partial^{a\bullet b}_{1\bullet\varepsilon} \ , \ \partial^{a\bullet b}_{1\bullet\varepsilon} \ \div^{s\bullet b}_{0\bullet 1}\ )$
$(\ \div^a_0 \ \partial^a_1 \ , \ \div^a_1 \ \partial^a_0\ )$	$(\ \div^{a\bullet b}_{0\bullet 1} \ \partial^{a\bullet b}_{1\bullet\varepsilon} \ , \ \div^{a\bullet b}_{1\bullet 1} \ \partial^{a\bullet b}_{0\bullet\varepsilon}\ )$

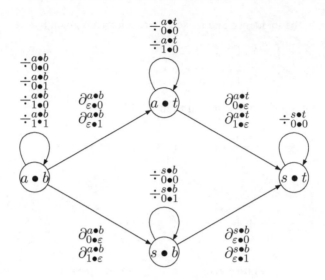

Figure 4.14. *Automaton of a surface automatically generated from the automatons of the two curves*

Figure 4.15 shows the quotient graph that we obtain from these six equations.

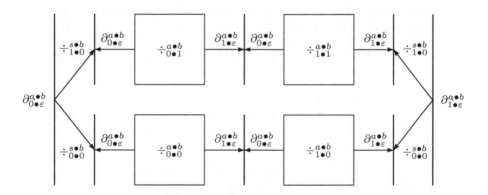

Figure 4.15. *Quotient graph of a surface induced by adjacency relations combining $\gamma_{\div}^a$ with $\Sigma_{\div}^b$*

Adjacency relations are completed by the other direction, that is the combination of $\Sigma_{\div}^{a}$ with $\gamma_{\div}^{b}$:

	$\div_{0^a}$			
$(\ \div_{0^b}\ \partial_{0^b}\ ,\ \partial_{0^b}\ \div_{0^t}\ )$	$(\ \div_{0^a\bullet 0^b}\ \partial_{\varepsilon^a\bullet 0^b}\ ,$	$\partial_{\varepsilon^a\bullet 0^b}\ \div_{0^a\bullet 0^t}\ )$
$(\ \div_{1^b}\ \partial_{1^b}\ ,\ \partial_{1^b}\ \div_{0^t}\ )$	$(\ \div_{0^a\bullet 1^b}\ \partial_{\varepsilon^a\bullet 1^h}\ ,$	$\partial_{\varepsilon^n\bullet 1^n}\ \div_{0^n\bullet 0^t}\ )$
$(\ \div_{0^b}\ \partial_{1^b}\ ,\ \div_{1^b}\ \partial_{0^b}\ )$	$(\ \div_{0^a\bullet 0^b}\ \partial_{\varepsilon^a\bullet 1^b}\ ,$	$\div_{0^a\bullet 1^b}\ \partial_{\varepsilon^a\bullet 0^b}\ )$

	$\div_{1^a}$			
$(\ \div_{0^b}\ \partial_{0^b}\ ,\ \partial_{0^b}\ \div_{0^t}\ )$	$(\ \div_{1^a\bullet 0^b}\ \partial_{\varepsilon^a\bullet 0^b}\ ,$	$\partial_{\varepsilon^a\bullet 0^b}\ \div_{1^a\bullet 0^t}\ )$
$(\ \div_{1^b}\ \partial_{1^b}\ ,\ \partial_{1^b}\ \div_{0^t}\ )$	$(\ \div_{1^a\bullet 1^b}\ \partial_{\varepsilon^a\bullet 1^b}\ ,$	$\partial_{\varepsilon^a\bullet 1^b}\ \div_{1^a\bullet 0^t}\ )$
$(\ \div_{0^b}\ \partial_{1^b}\ ,\ \div_{1^b}\ \partial_{0^b}\ )$	$(\ \div_{1^a\bullet 0^b}\ \partial_{\varepsilon^a\bullet 1^b}\ ,$	$\div_{1^a\bullet 1^b}\ \partial_{\varepsilon^a\bullet 0^t}\ )$

It remains to generate $\gamma_{\div}^{a\bullet t}$ by combining $\gamma_{\div}^{a}$ with $\Sigma_{\div}^{t}$:

	$\div_{0}^{t}$			
$(\ \div_{0}^{a}\ \partial_{0}^{a}\ ,\ \partial_{0}^{a}\ \div_{0}^{s}\ )$	$(\ \div_{0\bullet 0}^{a\bullet t}\ \partial_{0\bullet\varepsilon}^{a\bullet t}\ ,$	$\partial_{0\bullet\varepsilon}^{a\bullet t}\ \div_{0\bullet 0}^{s\bullet t}\ )$
$(\ \div_{1}^{a}\ \partial_{1}^{a}\ ,\ \partial_{1}^{a}\ \div_{0}^{s}\ )$	$(\ \div_{1\bullet 0}^{a\bullet t}\ \partial_{1\bullet\varepsilon}^{a\bullet t}\ ,$	$\partial_{1\bullet\varepsilon}^{a\bullet t}\ \div_{0\bullet 0}^{s\bullet t}\ )$
$(\ \div_{0}^{a}\ \partial_{1}^{a}\ ,\ \div_{1}^{a}\ \partial_{0}^{a}\ )$	$(\ \div_{0\bullet 0}^{a\bullet t}\ \partial_{1\bullet\varepsilon}^{a\bullet t}\ ,$	$\div_{1\bullet 0}^{a\bullet t}\ \partial_{0\bullet\varepsilon}^{a\bullet t}\ )$

and $\gamma_{\div}^{s\bullet b}$, and combining $\Sigma_{\div}^{s}$ with $\gamma_{\div}^{b}$:

	$\div_{0}^{s}$			
$(\ \div_{0}^{b}\ \partial_{0}^{b}\ ,\ \partial_{0}^{b}\ \div_{0}^{t}\ )$	$(\ \div_{0\bullet 0}^{s\bullet b}\ \partial_{\varepsilon\bullet 0}^{s\bullet b}\ ,$	$\partial_{\varepsilon\bullet 0}^{s\bullet b}\ \div_{0\bullet 0}^{s\bullet t}\ )$
$(\ \div_{1}^{a}\ \partial_{1}^{b}\ ,\ \partial_{1}^{b}\ \div_{0}^{t}\ )$	$(\ \div_{0\bullet 1}^{s\bullet b}\ \partial_{\varepsilon\bullet 1}^{s\bullet b}\ ,$	$\partial_{\varepsilon\bullet 1}^{s\bullet b}\ \div_{0\bullet 0}^{s\bullet t}\ )$
$(\ \div_{0}^{b}\ \partial_{1}^{b}\ ,\ \div_{1}^{b}\ \partial_{0}^{b}\ )$	$(\ \div_{0\bullet 0}^{s\bullet b}\ \partial_{\varepsilon\bullet 1}^{s\bullet b}\ ,$	$\div_{0\bullet 1}^{s\bullet b}\ \partial_{\varepsilon\bullet 0}^{s\bullet b}\ )$

4.2.2.3. *Adjacency relations on incidence operators*

We generate the elements of $\gamma_{\partial}^{a\bullet b}$ from equations [4.12]. In our case, we only use the third term of the equation since $\gamma_{\partial}^{a} = \gamma_{\partial}^{b} = \emptyset$. We combine the elements of Σ_{∂}^{a} and Σ_{∂}^{b}:

	∂_{0^b}	
∂_{0^a}	$(\ \partial_{0^a\bullet\varepsilon^b}\ \partial_{\varepsilon^s\bullet 0^b}\ ,$	$\partial_{\varepsilon^a\bullet 0^b}\ \partial_{0^a\bullet\varepsilon^t}\ )$
∂_{1^a}	$(\ \partial_{1^a\bullet\varepsilon^b}\ \partial_{\varepsilon^s\bullet 0^b}\ ,$	$\partial_{\varepsilon^a\bullet 0^b}\ \partial_{1^a\bullet\varepsilon^t}\ )$

	∂_{1^b}	
∂_{0^a}	$(\ \partial_{0^a\bullet\varepsilon^b}\ \partial_{\varepsilon^s\bullet 1^b}\ ,$	$\partial_{\varepsilon^a\bullet 1^b}\ \partial_{0^a\bullet\varepsilon^t}\ )$
∂_{1^a}	$(\ \partial_{1^a\bullet\varepsilon^b}\ \partial_{\varepsilon^s\bullet 1^b}\ ,$	$\partial_{\varepsilon^a\bullet 1^b}\ \partial_{1^a\bullet\varepsilon^t}\ )$

$$\gamma_{\partial}^{a\bullet b} = \left\{ \begin{array}{l} (\ \partial_{0^a\bullet\varepsilon^b}\ \partial_{\varepsilon^s\bullet 0^b}\ ,\ \partial_{\varepsilon^a\bullet 0^b}\ \partial_{0^a\bullet\varepsilon^t}\ ) \\ (\ \partial_{0^a\bullet\varepsilon^b}\ \partial_{\varepsilon^s\bullet 1^b}\ ,\ \partial_{\varepsilon^a\bullet 1^b}\ \partial_{0^a\bullet\varepsilon^t}\ ) \\ (\ \partial_{1^a\bullet\varepsilon^b}\ \partial_{\varepsilon^s\bullet 0^b}\ ,\ \partial_{\varepsilon^a\bullet 0^b}\ \partial_{1^a\bullet\varepsilon^t}\ ) \\ (\ \partial_{1^a\bullet\varepsilon^b}\ \partial_{\varepsilon^s\bullet 1^b}\ ,\ \partial_{\varepsilon^a\bullet 1^b}\ \partial_{1^a\bullet\varepsilon^t}\ ) \end{array} \right\}$$

These equations define the cellular decomposition of the face into faces/edges/vertices, with identification of the vertices common to the different edges, as illustrated in Figure 4.16.

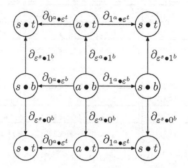

Figure 4.16. *Quotient graph obtained from adjacency relations on incidence operators for a surface*

4.2.3. *Volume generation from curves*

The generation of volumes or even of structures in higher dimensions is immediate due to the associativity property of the combination operator $\bullet$. The product $a \bullet b \bullet c$ (where a, b and c represent three curves) is determined by calculating $(a \bullet b) \bullet c$. Figure 4.17 shows the automaton thus built.

4.2.4. *Generation of the tree structure*

The idea is to define an operator to automatically generate a BC-IFS which represents a tree whose leaves will correspond to a structure imposed and described by a BC-IFS. An example is to automatically generate a tree structure that supports a NURBS surface, which could be used in architecture or additive manufacturing to create overhangs while saving material.

The principle consists of building a tree connecting a vertex s described by the BC-IFS $(Q_1 = \{s\}, \Sigma_1 = \{0^s\}, \delta_1 = \{((s, 0^s), s)\}, \gamma_1 = \{\})$ and a figure b described by a second BC-IFS $(Q_2, \Sigma_2, \delta_2, \gamma_2)$. As we have seen in section 2.5, and based on the example in Figure 4.20, we present the construction principle in three stages.

1) We define two edges, that is, two topological cells, each represented by an automaton state. The first is a vertex type that represents the root of the tree. The second, denoted by b, represents all the leaves and can be of different types. For example, in Figure 4.20 it is a curve.

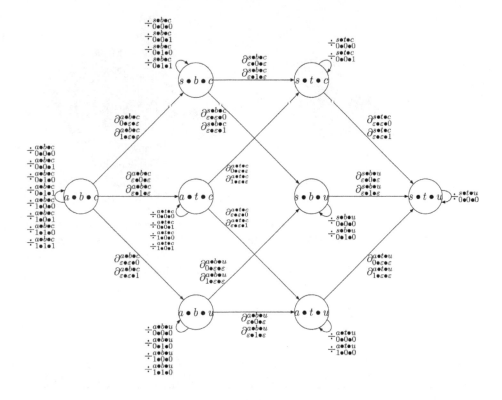

Figure 4.17. *Automaton of a volume structure obtained by tensor product of three BC-IFS representing curves*

2) Since the tree needs to "adjust" to all leaves, its subdivision process must be compatible with that of b. For each subdivision of the set of leaves, we define a subdivision of the tree into a branch and a tree. Each branch must be connected from one side to the initial root, and from the other to the root of the new tree. The new tree must be connected to the corresponding subdivision in the set of leaves.

3) If the set of leaves b is subdivided into other types of cells other than itself, the tree must be subdivided into other types of trees, as shown in Figure 4.20. We need to define as many types of trees as states of Q_2 that are accessible from b by subdivision.

Let us formally describe the method, specifying the construction: of the set of states, of the set of symbols, transitions, and incidence and adjacency relations.

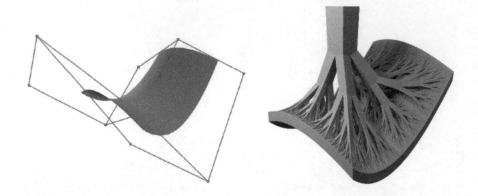

Figure 4.18. *Example of an automatically generated tree, whose leaves will perfectly match a structure described by a BC-IFS: on the left, a quadratic NURBS surface; on the right, a tree structure whose branches are delimited by the imposed surface. For a color version of this figure, see www.iste.co.uk/gentil/geometric.zip*

Figure 4.19. *First stages of subdividing a tree bordered by a curve*

We denote by $\triangleleft$ the operator generating the BC-IFS representing the final tree, from its two borders:

– the root described by $(Q_1 = \{s\}, \Sigma_1 = \{0^s\}, \delta_1 = \{((s, 0^s), s)\}, \gamma_1 = \{\})$;

– the set of leaves, described by the BC-IFS $(Q_2, \Sigma_2, \delta_2, \gamma_2)$.

4.2.4.1. *States*

To define the entire cellular decomposition of the tree structure, the set of states Q must contain the states of each edge (Q_1 and Q_2), the new state a representing the branch, and as many states for the different types of trees as for the different types of states in which b is decomposed:

$$Q = \{s \triangleleft x \mid x \in \delta_2^*(b, L_{\div}^b)\} \cup Q_1 \cup Q_2 \cup \{a\} \qquad [4.13]$$

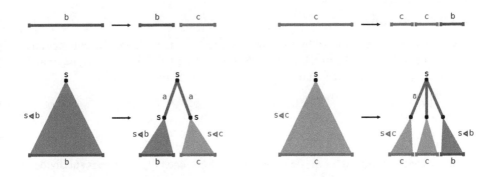

Figure 4.20. *The types of subdivisions of the tree depend on the types of edge subdivisions. For a color version of this figure, see www.iste.co.uk/gentil/geometric.zip*

4.2.4.2. *Symbols*

We need to define two incidence operators for each type of tree: one for the root that we denote $\partial_{0^{s \lhd x}}$ and one for the set of leaves, denoted as $\partial_{1^{s \lhd x}}$:

$$\forall x \in \delta_2^*(b, L_{\div}^b) \; ; \; \Sigma_{\partial}^{s \lhd x} = \{\partial_0^{s \lhd x}, \partial_1^{s \lhd x}\} \tag{4.14}$$

Each type of tree is divided into a branch and a tree. We define subdivision symbols for each tree type based on the subdivision symbols of each subset of leaves that we denote $\div_{u_\lhd}^{s \lhd x}$ for their subdivision into a tree and $\div_{u_-}^{s \lhd x}$ for their subdivision into a branch:

$$\forall x \in \delta_2^*(b, L_{\div}^b)$$
$$\Sigma_{\div}^{s \lhd x} = \{\div_{u_-}^{s \lhd x} \mid \div_u^x \in \Sigma_{\div}^x\} \cup \{\div_{u_\lhd}^{s \lhd x} \mid \div_u^x \in \Sigma_{\div}^x\} \tag{4.15}$$

Finally, we need to define the symbols for the branch a:

$$\Sigma_{\div}^a = \{\div_0^a\} \; , \; \Sigma_{\partial}^a = \{\partial_0^a, \partial_1^a\} \tag{4.16}$$

REMARK.– Here, we have chosen to follow the example shown in Figure 4.20 by defining the branch as being a segment. We have defined only one subdivision operator that will have to be the identity operator. The branch is then defined by the display primitive associated with the state. A contractive operator would have reduced the branch to one point. There are endless other possibilities available to us. For example, we could have defined a curve with two or three subdivision operators.

4.2.4.3. *Transitions*

Branch a must be bordered by two vertices s to be able to define the connections with the subdivision of the initial root on the one hand (which is also of type s) and with the vertex of the subtree on the other hand. The transitions are then:

$$\delta(a, \div_0^a) = a$$
$$\delta(a, \partial_0^a) = s \qquad\qquad [4.17]$$
$$\delta(a, \partial_1^a) = s$$

As we have mentioned for the definition of symbols, each type of tree is subdivided into a subtree (taking the subdivisions of b into account) and a branch a:

$$\forall x \in \delta_2^*(b, L_\div^b) \; ; \; \forall \div_u^x \in \Sigma_\div^x$$
$$\delta(s \triangleleft x, \div_{u_-}^{s \triangleleft x}) = a \qquad\qquad [4.18]$$
$$\delta(s \triangleleft x, \div_{u_\triangleleft}^{s \triangleleft x}) = s \triangleleft \delta_2(x, \div_u^x)$$

The special case in Figure 4.20 generates the automaton presented by Figure 4.21.

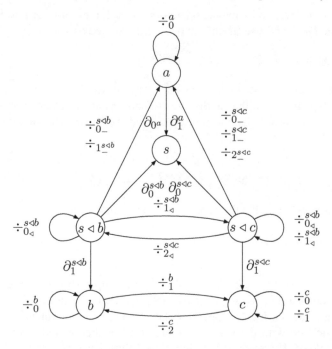

Figure 4.21. *Automaton representing a tree structure whose set of leaves is divided into two different types of curves as presented by Figure 4.20 (the left diagram). This automaton was generated automatically from the BC-IFS to write the set of leaves*

4.2.4.4. *Equivalence relations*

Each tree is divided into a branch and a tree. These two subdivisions must be connected to each other via a vertex, hence the equivalence relations such as $(\div_{u_-}^{s\triangleleft x} \partial_1^a , \div_{u_\triangleleft}^{s\triangleleft\delta(x,\div_u^x)} \partial_0^{s\triangleleft x})$. The branch must be connected on the other side to the root subdivision by means of the equivalence relations such as $(\partial_0^{s\triangleleft x} \div_0^s ; \div_{u_-}^{s\triangleleft x} \partial_0^a)$. Finally, the tree must be connected by its other edge (of the curve type in our example) to a subdivision of the set of leaves $(\div_{u_\triangleleft}^{s\triangleleft x} \partial_1^{s\triangleleft\delta(x,\div_u x)} ; \partial_1^{s\triangleleft x} \div_u^x)$, which in the end yields the following set of equivalence relations:

$$
\begin{aligned}
\forall x &\in \delta_2^*(b, L_\div^b) \\
\gamma_\div^{s\triangleleft x} &= \{(\partial_0^{s\triangleleft x} \div_0^s , \div_{u_-}^{s\triangleleft x} \partial_0^a) \mid \div_u^x \in \Sigma_\div^x\} \cup \\
&\quad \{(\div_{u_-}^{s\triangleleft x} \partial_1^a , \div_{u_\triangleleft}^{s\triangleleft\delta(x,\div_u^x)} \partial_0^{s\triangleleft x}) \mid \div_u^x \in \Sigma_\div^x\} \cup \\
&\quad \{(\div_{u_\triangleleft}^{s\triangleleft x} \partial_1^{s\triangleleft\delta(x,\div_u x)} ; \partial_1^{s\triangleleft x} \div_u^x) \mid \div_u^x \in \Sigma_\div^x\}
\end{aligned}
\tag{4.19}
$$

All that remains is to define the incidence relations for the branches a:

$$
\gamma_\div^a = \{(0\ \partial_0 , \partial_0\ 0) ; (0\ \partial_1 , \partial_1\ 0)\}
\tag{4.20}
$$

Figure 4.22. *Double trees bordered by quadrangular surfaces*

4.3. Applications

4.3.1. *Interaction between representation models*

In Chapter 3, we have shown that with BC-IFS it is possible to equally represent NURBS, subdivision surfaces as well as fractals, naturally. One of the consequences is to be able to mix and make these different types of objects interact using the same model. For example, using the tensor product, we can build a surface from a NURBS curve and a fractal curve to get a smooth surface in one direction and a rough surface in the other direction (see Figure 4.23).

Figure 4.23. *Example of surface obtained by the tensor product of a NURBS curve with a fractal curve. The resulting surface presents a smooth appearance following one direction and a rough one in the other direction. For a color version of this figure, see www.iste.co.uk/gentil/geometric.zip*

Another application builds tree structures that support or consolidate an imposed NURBS surface. Construction is carried out exactly according to the method described in section 2.5 and automated in section 4.2.4. The NURBS surface is considered to be an imposed boundary surface and the tree is designed to grow to that boundary surface (see Figure 4.18).

4.3.1.1. *Primal scheme versus dual scheme and BC-IFS*

Another relevant application that emphasizes the relevance of the BC-IFS model is the construction of connections between surfaces of different types. We present the case of the construction of a connection between a dual subdivision scheme (the Doo–Sabin scheme for a quadrangular face (A) and a primal scheme (the Catmull–Clark scheme for a quadrangular face (B). One of the issues for this type of problem is to address it based on control meshes, that is to find a control mesh configuration, connecting the two initial meshes. This approach raises difficulties because, during the subdivision, the primal scheme "preserves" mesh points, while the dual scheme does not retain any. However, if we reason at the topological level, these two patterns operate in the same way, by subdividing the quadrangular faces into four parts. The real problem is connecting the surfaces and not the control meshes. The interest of the BC-IFS is to code the topology of the limit surface. It is thus possible to define connecting constraints directly on limit surfaces. The other advantage is to have a display primitive that makes it possible to approximate the limit surface at each iteration level. This primitive can be chosen so as to have a tesselation of the limit surface.

To build the connection, we define in the center an intermediate structure J achieving the junction between faces A and B found on either side (see Figure 4.24). This intermediate structure connects on one side to face A via an edge. The intermediate structure must at least share the same control points as those of the edge of face A. The same holds for face B. The dimension of the barycentric space associated with face J must be at least equal to the dimension of the edge of face A added to that of the space of the edge of face B. Then we can add as many internal dimensions as we want to increase the control of the geometry of J. The principle is then to subdivide J into six parts: two of type A on the side of face A, two of type B on the side of face B and two of type J in the center. Incidence and adjacency constraints are naturally written following the subdivision schemes. Figure 4.25 highlights the behavior of the subdivision process. This illustration shows that the attractor associated with the state J (in green in the figure) is a curve. Indeed, there are only two transitions ending at state J and thus defining a curve. This subdivision process extends each face (A and B) in the direction of the other via copies of itself. The connection of these copies is controlled by the incidence and adjacency constraints. The intermediate structure J helps guide this construction. This method is general and can be used for all types of surface pairs. One variant is to use two types of intermediate structures: one from each side to independently control the copies of each face. These intermediate structures are connected to each other. Figure 4.26 shows different examples.

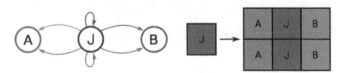

Figure 4.24. *Automaton representing the construction of a connection structure between two different types of faces (types A and B). Faces A and B are quadrangular faces and are subdivided into four. For a color version of this figure, see www.iste.co.uk/gentil/geometric.zip*

4.3.2. *Structure optimization*

The examples of Chapter 2 show the variety of forms that we can generate using an iterative process. One of the main reasons for this variety is the ability to simply define new forms of complex topologies (lacunar or tree-based). The other contribution of this approach is the ability to produce geometries that are "perturbed", "chaotic" or rough. The multiplication of patterns and details at different iteration levels gives an esthetic aspect to these forms. Figure 4.27 shows the example of a shading panel built from the model of a lacunar surface.

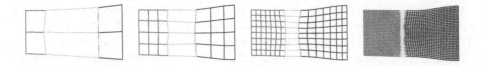

Figure 4.25. *Connection subdivision process. The red, green and blue mesh, respectively, represents the primitives of the faces of types* A, J *and* B. *At each iteration, the green faces are replaced by red and blue faces. For a color version of this figure, see www.iste.co.uk/gentil/geometric.zip*

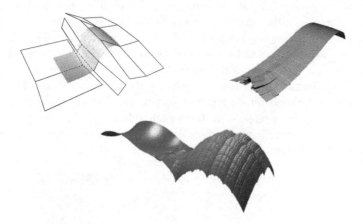

Figure 4.26. *Connections built between different pairs of surfaces: a Doo–Sabin surface and a Catmull–Clark surface; a Catmull–Clark surface and a surface obtained by the tensor product of two von Koch curves; a Catmull–Clark surface and a "Takagi" surface, obtained by the tensor product of two "Takagi" curves. For a color version of this figure, see www.iste.co.uk/gentil/geometric.zip*

These are geometries that we can frequently encounter in nature, but which until now have been difficult to represent and are really difficult or impossible to manufacture. This is certainly one of the reasons why few geometric patterns have been developed to represent them.

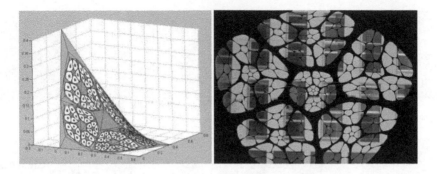

Figure 4.27. *Example of a lacunar surface, co-imagined by architects (IBOIS-EPFL) and computer scientists (LIRIS), to design a shading panel. The geometric model is represented on the left, and on the right the lacunar surface is obtained by slicing. For a color version of this figure, see www.iste.co.uk/gentil/geometric.zip*

The representation model outlined in this book is based on the self-similarity property. One of the great difficulties in designing is to imagine a new object by defining it using a self-reference. This is a new way of thinking about the design approach. Here are some examples of these steps.

4.3.2.1. *Design of lacunar forms*

The design of lacunar shapes can be done by following two approaches. The first consists of performing a self-similar tiling of the space, and then of removing elements from the tiling, thus producing a lacuna that is reproduced at each iteration level. Classic examples are as follows: in 1D, the Cantor set; in 2D, the Sierpinski triangle and the Menger carpet; in 3D, the Menger sponge. The second approach is based on the composition of elements. From basic elements, we compose a structure, so as to leave one or more lacunas, which multiply over the iterations. Unlike the previous method, this construction is not constrained by a predefined tiling and offers more freedom.

4.3.2.1.1. Design by deletion of elements

The Cantor set can be defined as the slicing of the segment $[0, 1]$ into three segments, from which we eliminate the middle segment. The operation is iterated over the remaining two segments. Similarly, the Sierpinski triangle can be constructed by breaking down a triangle into four triangles and then eliminating the central triangle. The Menger carpet is obtained from the subdivision of a square into nine squares and also by removing the central square (see Figure 4.28).

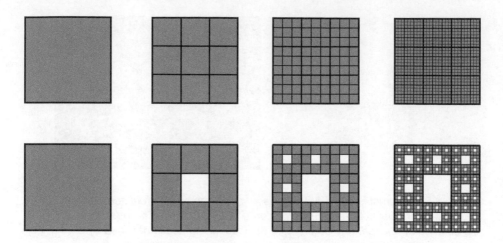

Figure 4.28. *Example of the definition of a 2D lacunar structure
by tiling the plane and then removing elements*

COMMENTS ON FIGURE 4.28.– *Here, the Menger sponge can be seen as an incomplete space tiling: the square is subdivided into nine parts and the central part is removed, leaving a lacuna. As the subdivisions progress, the lacuna is duplicated into each sub-part, increasing the lacunarity of the structure.*

In 3D, the Menger sponge is built from the subdivision of a cube into 27 parts of which seven cubes are eliminated (one in the center of the cube and six located in the center of each face, see Figure 4.29). The corresponding topological structure can be quickly defined by creating the tensor product of three curves subdivided into three parts. Then, one simply has to eliminate the seven subdivision operators corresponding to the green cubes as well as the incidence and adjacency relations in which they are involved. The final structure can then be distorted using the control and subdivision points while maintaining the topology, as illustrated by Figure 2.34.

Similar examples can be constructed from the pentagonal or hexagonal subdivisions seen in section 2.3.3 (see Figure 4.30).

This approach can be applied in a variety of ways. For example, we can increase the number of elements defining the regular tiling. In this way, we are also increasing the possibilities of creating lacunas. However, these constructions are relatively limited because of the geometric characteristics of the structures that we can produce. Indeed, a structure such as the Menger sponge is strongly linked to the topological structure of the tensor product type, from which it was built. For example, if we want to round the corners of this structure, we can do so by choosing an evaluation primitive that is better suited. However, as we have seen, the final attractor is independent of this primitive, and as iterations proceed, the form is inexorably approaching its attractor, which has sharp angles. If we use polynomial basis functions to distort the faces of the sponge, we shall only be able to influence the overall geometry, and locally the structure will present sharp angles, as shown in Figure 2.48.

Figure 4.29. *Example of construction of the Menger sponge. In red, the cube is decomposed into 27 cubes. The seven green cubes are removed to create gaps in the blue structure. For a color version of this figure, see www.iste.co.uk/gentil/geometric.zip*

Figure 4.30. *Lacunar structures obtained by removing the central part of a regular subdivision: pentagonal on the left (straight edges), hexagonal at the center (curved edges) and hexagonal on the right (the edges are Cantor sets). For a color version of this figure, see www.iste.co.uk/gentil/geometric.zip*

A variant of this approach is to use space tilings with several types of tiles. The principle consists of having a set of tiles and to subdivide each of these tiles using a subset of this family of tiles. The use of several types of tiles increases the possibilities of space decomposition and creation. We present an example which we call the Menger–Excoffier carpet. The initial problem, submitted by Thierry Excoffier, was to avoid having a non-homogeneous distribution of lacunas in the Menger carpet and sponge. The idea is to build a similar structure (the Menger–Excoffier carpet and sponge) but whilst exhibiting greater homogeneity, so that the walls between the lacunas are of the same thickness, as presented by Figure 4.31. The 2D problem corresponds to a face of the Menger–Excoffier sponge. To define this structure, we use two types of tiles. The first is subdivided as the standard Menger carpet, but by involving the other type of tile (seen in red in Figure 4.32). This second tile is subdivided into an *H* consisting of the two types of tiles.

Figure 4.31. *Menger–Excoffier sponge with walls of the same thickness between each lacuna (source: project MODITERE no. ANR-09-COSI-014). For a color version of this figure, see www.iste.co.uk/gentil/geometric.zip*

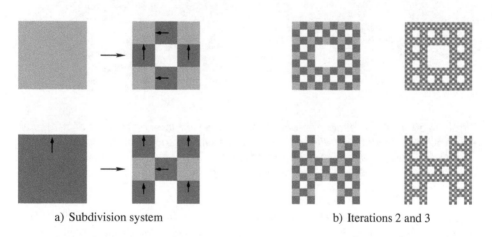

a) Subdivision system   b) Iterations 2 and 3

Figure 4.32. *The Menger–Excoffier sponge is built from two subdivision systems of the square. For a color version of this figure, see www.iste.co.uk/gentil/geometric.zip*

COMMENTS ON FIGURE 4.32.– *The first subdivision system is applied to the blue elements and is represented by the top diagram 4.32(a). The square is subdivided into nine tiles and the central tile is removed. The tiles, located at the four corners of the square, follow the same subdivision system (blue), while the others follow the second one (red). The second subdivision system is applied to the red tiles and is presented by the bottom diagram 4.32(a). The square is also subdivided into nine tiles but this time the two central tiles of the first and last row are the ones removed. The central pavers of the first and last columns follow the subdivision rule of the blue tiles and the others follow that of the red tiles. An arrow indicates the orientation of the red tiles for which the subdivision is oriented. Diagram 4.32(b) shows iterations 2 and 3.*

Figure 4.33. *For the Menger–Excoffier sponge, the two 3D subdivision systems are similar to the 2D rules: on the left, the blue cube subdivision; on the right, the red cube. For a color version of this figure, see www.iste.co.uk/gentil/ geometric.zip*

4.3.2.1.2. Design by composition of elements

The design of shapes by composition of elements is more general, and allows for generating a more significant variety of forms, but the approach also proves to be more complex. It can be achieved using one or more types of cells. For each of these cells, representing any topological structure, we must define its cellular decomposition. These cellular decompositions help to define the edges at which we shall connect them to define the subdivision system and, in our case, to define lacunas. One of the difficulties is building a consistent system in order for the subdivisions of the borders to correspond to borders of subdivision, including the additional orientation constraint described in section 4.1. These can be lifted at the cost of additional work consisting of introducing permutation operators in the incidence and adjacency constraints. Figure 4.3 illustrates this approach. Lacunas are defined in the center of each hexagonal cell in the form of a star. If we want to mitigate the angular zones induced by the branches of these stars, it is possible to modify the geometry using subdivision points. However, this type of modification can cause non-homogeneity of contraction directions and thus crush cells and lacunas in the most contractive direction. An alternative is to fill the branches by adding triangular cells, following the pattern shown in Figure 4.34.

We present two 3D examples, illustrating this construction principle.

EXAMPLE.– For this first example, the starting cell is a tetrahedron. The construction is similar to that of the triangle (previously described). For the 2D construction, the triangle is "sliced" at each vertex to obtain: three smaller triangles and a hexagon at the center (which is a truncated triangle). Starting with a tetrahedron, we can divide it into four smaller tetrahedrons (at each of the vertices) and a truncated tetrahedron (Figure 4.35 on the left). The truncated tetrahedron is subdivided into 12 truncated tetrahedrons (red) and 12 tetrahedrons (blue). In this example, one out of two of the branches were filled for the stars of the faces, so as not to reduce the lacunarity too much. Two examples of geometric embedding are shown in Figure 4.36 and two examples of prototypes designed via the assembly of these tetrahedral structures are presented by Figure 4.37.

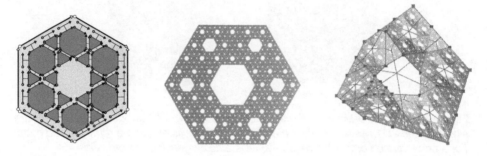

Figure 4.34. *"Triangle"-type cells have been added to the structure of Figure 4.3 to define a convex lacuna: on the left, the subdivision scheme with incidence and adjacency relations; in the center, the flat structure represented at iteration level 5; on the right, the same surface structure with geometric embedding in $\mathbb{R}^3$. The subdivision of the triangular cell remains the same as shown in Figure 4.3. For a color version of this figure, see www.iste.co.uk/gentil/geometric.zip*

Figure 4.35. *Design of a lacunar structure from two types of cell: a tetrahedron (in blue) and a truncated tetrahedron (in red). For a color version of this figure, see www.iste.co.uk/gentil/geometric.zip*

COMMENTS ON FIGURE 4.35.– *On the left, the tetrahedron is subdivided into four tetrahedrons (positioned at each of the vertices of the initial tetrahedron) and a truncated tetrahedron (positioned at the center) to recompose the full initial tetrahedron. In the center, the truncated tetrahedron is subdivided into 12 truncated tetrahedrons (positioned at each of the vertices of the initial truncated tetrahedron), revealing lacunas. Parts of the lacunas are filled up with 12 tetrahedrons to make the structure rigid. On the right the result of the application of two iterations of the subdivision process is outlined.*

Figure 4.36. *Two copies of the 3D lacunar topological structure obtained from the subdivision rules of the scheme shown in Figure 4.35: on the left side, the geometric embedding is close to the schematic representation of the cells; on the right side, it has been modified using subdivision points. For a color version of this figure, see www.iste.co.uk/gentil/geometric.zip*

Figure 4.37. *Prototypes designed by assembly of tetrahedral structures shown in Figure 4.36 (source: project MODITERE no. ANR-09-COSI-014). For a color version of this figure, see www.iste.co.uk/gentil/geometric.zip*

With this example, we observe that despite topological constraints, we can produce very different geometries, but only by manipulating the subdivision points (that is, the local geometry). The shape of the lacunas strongly influences the appearance of the structure and both its geometric and physical characteristics (density, mass, etc.).

EXAMPLE.– With this second example, we illustrate a slightly more elaborated approach to the construction of porous structures. We start by illustrating the principle of our approach with a surface structure. The idea is to start from a classical topological cell: here a hexagonal face (see Figure 4.38). This face is bordered by two types of edges: three red and three blue, arranged alternately, around the face. Blue edges are used to border the lacunas (that is, to define the boundary between matter and void) and red edges are used to connect cell subdivisions to each other. However, these red edges will be Cantor set-type edges to allow a lacuna to be integrated at the level of every connection. Figure 4.39 shows the result of this construction for iterations 3 and 4.

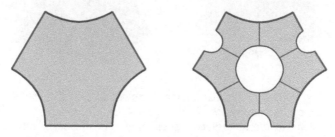

Figure 4.38. *Illustration of the topological subdivision of the lacunar face. For a color version of this figure, see www.iste.co.uk/gentil/geometric.zip*

Figure 4.39. *Lacunar face at iteration levels 3 (on the left) and 4 (in the center). On the right, geometric embedding with the network of control points (in red) and subdivision points (in blue). For a color version of this figure, see www.iste.co.uk/gentil/geometric.zip*

In Figure 4.40, we can see more precisely how cellular decomposition is organized at the face and edge levels. The incidence relations are defined for each edge (red as blue). Adjacency relations should only be defined between the red edges. In this configuration, we can see that, in order to achieve the connections between the subdivided cells, the orientations of the red edges are not compatible. We must therefore use a permutation operator to write the adjacency relations.

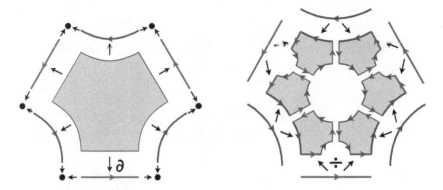

Figure 4.40. *Illustration of the cellular decomposition and the subdivision process of the face and edges. For a color version of this figure, see www.iste.co.uk/gentil/geometric.zip*

The same construction can be carried out in 3D. Let us take a truncated tetrahedron as a starting cell (see Figure 4.41). The triangular faces originating from the truncating (in red in the figure) will be used to carry out the connections between the subdivided cells. The other faces (in blue) will be of the same nature as the previously built lacunar face, and will be at the interface between matter and void. Figure 4.41 shows iterations 1–5.

4.3.2.2. *Filling by porous volumes*

As we have just seen, the topologies and geometries that we can generate using BC-IFS are remarkable and seem to be adapted to design lightweight structures. One question that comes up often is: is it possible to "fractalize" any kind of object? Or in other words: given a geometric shape, is it possible to replace it with a lacunar structure in order to lighten it?

We can address this issue in a number of ways. The first consists of following one of the steps we have presented in section 4.3.2.1. That is, consider this form as an initial cell that we will be duplicating into identical smaller cells (or eventually using other types of cells) and connect them in order to restore this initial form. For some basic forms, this may be possible, as we have shown with the examples of this same section. However, for more complex forms, such as those of a mechanical system or, for example, an elephant (see Figure 4.42), the implementation can become very complex if not impossible. As such, a simple solution is to achieve a tiling of this shape (tetrahedralization, hexahedralization) and replace every tile with a lacunar version of the same tile. For Figure 4.36, we have performed an assembly of lacunar tetrahedrons based on a tetrahedral decomposition of the elephant.

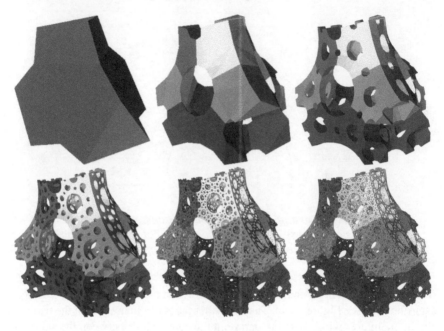

Figure 4.41. *Example of the design of a lacunar structure, built from a truncated tetrahedron. For a color version of this figure, see www.iste.co.uk/gentil/geometric.zip*

COMMENTS ON FIGURE 4.41.– *The red faces are used to make the connections between the subdivided cells. These faces have a "triangular Cantor" structure where a triangle is subdivided into three unconnected triangles. The blue faces, due to the interaction of volume connections, are identical to faces shown in Figure 4.39, given as a 2D example.*

Another solution is to define a basic block and build assemblies of that block. Figure 4.43 shows two assemblies constructed from the truncated tetrahedrons of Figure 4.41. The connections to the pedestals (located at the top and bottom of the structures) have been made using tree-based structures whose trunks are also composed of truncated tetrahedrons.

For Figure 4.44, the assembly has been achieved using a diamond-like structure. Each truncated tetrahedron corresponds to an atom, and the truncated faces correspond to the bonds between atoms.

Figure 4.42. *Example of filling using porous volumes. On the left, the geometry of an elephant described by a mesh (source: CGAL). In the center and on the right, respectively, the volume described by the envelope of the shape has been tetrahedralized and each tetrahedron has been replaced by one of the lacunar tetrahedrons of Figure 4.36. For a color version of this figure, see www.iste.co.uk/gentil/geometric.zip*

Figure 4.43. *Examples of the design of lacunar structures through the assembly of lacunar blocs, presented in Figure 4.41. The connections with the pedestals have been achieved with the tree-based design method in which trunks are themselves lacunar blocks. For a color version of this figure, see www.iste.co.uk/gentil/geometric.zip*

For structures achieved through the assembly of lacunar truncated tetrahedrons, the material gain is 57% at every iteration. In order for the structures to be effectively manufacturable via 3D printing, the reasonable number of iterations is in the order of

4, exceptionally 5. Beyond this limit, the structures may be too thin to be materialized. However, with only four iterations, the material gain is 88%.

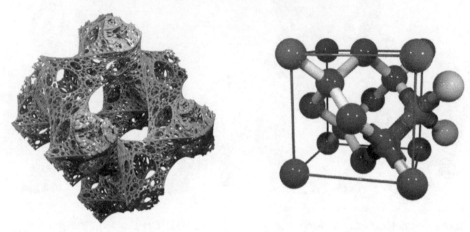

Figure 4.44. *Assembly according to the structure of a diamond. On the left, an example of a structure obtained via the assembly of lacunar blocks (here a truncated tetrahedron), organized according to a diamond-like structure. For a color version of this figure, see www.iste.co.uk/gentil/geometric.zip*

COMMENTS ON FIGURE 4.44.– *On the right-hand side, the structure of the diamond; the red atoms delineate the periodicity of the structure. In green and purple, an atom and its four bonds with neighboring atoms are shown. These structures are replaced by a truncated tetrahedron centered on the atom and each truncated face is normal with respect to the direction of the bonds. We may note that the green and purple structures are out of phase by $\frac{2\pi}{3}$. This phase shift results in a twist of the lacunar tetrahedral structures.*

4.3.3. *Roughness and space refilling*

Rough surfaces are interesting for their properties. They make it possible to increase the surface area between two media and can thus promote the exchange of heat or other. They also have a greater acoustic absorption capacity than smooth surfaces (Sapoval *et al.* 1997) because of their localized natural modes that have been experimentally identified (Mbailassem *et al.* 2018).

With the BC-IFS model (as with every model generating fractals), the roughness is obtained naturally in the general case, while differentiability is a special case (Bensoudane *et al.* 2009; Podkorytov *et al.* 2012; Sokolov *et al.* 2012). Figure 4.45 shows two examples of roughness of quadrangular surfaces defined from randomly selected subdivision points. These surfaces have been defined as a quadrangular surface with a standard subdivision (see Figure 1.52) and internal dimensions equal to 2, for the vertices, edges and face.

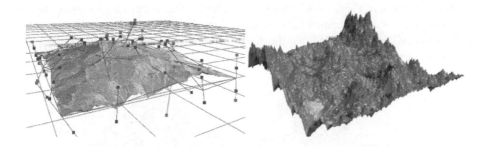

Figure 4.45. *Examples of rough surfaces. For a color version of this figure, see www.iste.co.uk/gentil/geometric.zip*

It is possible to increase the possibility of surface deformation by increasing the number of subdivisions, by mixing cell types, as we have seen for lacunarity, or by constituting "prominences" in a similar way to the von Koch curve (see Figure 2.6). The latter is constructed by subdividing a segment into three and replacing the middle segment with two segments forming an outgrowth in the original segment. This growth is propagated from subsegment to subsegment over the iterations. Depending on the shape given to this growth, it is possible to produce curves that almost fill up a surface (see Figure 4.46)

Figure 4.46. *Example of a variant of the Von Koch curve filling up almost an entire surface. From left to right, iterations 1, 2, 3 and 6 are represented. For a color version of this figure, see www.iste.co.uk/gentil/geometric.zip*

This principle can be used with surface structures to design rough surfaces. We have already mentioned such an example in section 2.6. Starting with a triangular face, with a subdivision identical to that of the Sierpinski triangle, we can fill the hole with a triangular-based pyramid to define this growth. We then obtain the surface as shown in Figure 2.45 and the resulting assembly is a 3D equivalent of the von Koch flake.

We can operate the same construction from the subdivision of a quadrangular surface into nine parts, as for the Menger carpet. But instead of removing the central part, we connect it with a cube with a missing face. Figure 4.47 shows iterations 1–4 of this construction. Figure 4.48 shows two variants of the same construction: for the image on the left, some details have been duplicated in hollows; for the image on the right, the face of the cube defining the prominence has rotated, inducing a twisting of the growth over the iterations.

Figure 4.47. *Example of a rough, or even chaotic, surface, designed on the principle of addition of an outgrowth. The illustrations correspond to iteration levels 1, 2, 3 and 4. For a color version of this figure, see www.iste.co.uk/gentil/geometric.zip*

Roughness can also be used to rigidify surfaces. According to the principle of origami, folds are like ribs rigidifying the structure. This principle was exploited by the IBOIS laboratory of the EPFL to build self-supporting hulls, while making use of the esthetic aspect of fractals. Figure 4.49 shows an example of a hull obtained by tensor product of a smooth curve and a fractal curve. The structure was manufactured by assembling wood panels (IBOIS EPFL-LIRIS).

4.3.4. *The design of a heat exchanger*

We present an example of an application designed using the MODITERE iterative modeler. The objective is to create a thermal exchanger using a fractal structure to increase the exchange area between two fluids and thus facilitate thermal exchanges. This structure is made up of three parts: two switches and an exchanger (see Figure 4.50). The exchanger is simply an extruded checkerboard. The switches are tree structures that distribute the fluids alternately in the squares of the checkerboard to increase the exchange area. These switches are placed at each end of the exchanger. The first "crosses" the fluids, and the second uncrosses them. In order for each of the two fluids to be alternately distributed into the squares of a checkerboard, the switch is designed as a tree whose boundary surface is a Hilbert/Peano curve (see section 2.3.1). At every iteration, the tree structure is extended with an additional floor, and each of these floors corresponds to an increasingly refined Hilbert/Peano curve. The trick then consists of using a display primitive linking a level 0 to a level 1 Hilbert/Peano curve. At level 0, the curve is a simple segment and at level 1 it consists of nine segments. One simply has to subdivide the level 0 segment into nine segments and join, in order, each of these nine segments to the other nine level 1 segments, as shown in Figure 4.51. At each iteration, the exchange area is multiplied by three.

Figure 4.48. *Both surfaces are designed from the same topological structure as shown in Figure 4.47 with, on the left-hand side, some growths recopied into hollows and, on the right-hand side, a twist induced by a rotation of the upper part of the outgrowth. For a color version of this figure, see www.iste.co.uk/gentil/geometric.zip*

Figure 4.49. *Example of a self-supporting hull, built by wood panel assembly (IBOIS EPFL-LIRIS). The surface was obtained by making the tensor product of a Bezier curve with a fractal curve. The final surface has been produced using three iterations (source: IBOIS EPFL-LIRIS). For a color version of this figure, see www.iste.co.uk/gentil/geometric.zip*

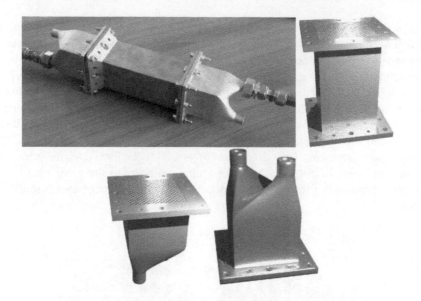

Figure 4.50. *On top, prototype of the thermal exchanger manufactured in aluminum by laser sintering (source: ©LIB). For a color version of this figure, see www.iste.co.uk/gentil/geometric.zip*

COMMENTS ON FIGURE 4.50.– *The exchanger is composed of three parts: a central part consisting of a simple extruded checkerboard and two switches, with on one side the inlet/outlet for each of the fluids and on the other the checkerboard structure. On one of the sides, the switch alternately distributes the two fluids into the checkerboard squares; on the other side, it gathers the fluids in each of the exit orifices.*

Figure 4.51. *Construction of the display primitive linking two levels of the exchanger tree. For a color version of this figure, see www.iste.co.uk/gentil/geometric.zip*

COMMENTS ON FIGURE 4.51.– *On the left, the segment is subdivided into nine segments. Each of these segments is extruded to connect one of the nine segments of the curve of the next level. Over the course of the iterations, this structure is duplicated from level to level. In the center and on the right side, the structure is similar but with a smoother geometry to facilitate fluid flows.*

Figure 4.52 shows different cross-sectional views achieved with the switch, pointing out the evolution of the tree structure.

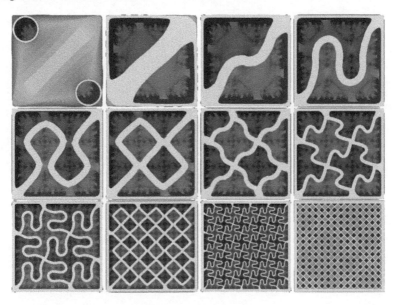

Figure 4.52. *Internal structure of the switch. For a color version of this figure, see www.iste.co.uk/gentil/geometric.zip*

COMMENTS ON FIGURE 4.52.– *Several cross-sections are made along the longitudinal axis to bring forward the structure of the switch. At the top left, the cross-section is made at the level of the two inlet/outlet of each fluid (transparent red and blue). Then, in order from left to right and from bottom to top: the first level shows the separation of the cavity in two; the second level shows the separation between the two cavities becoming distorted, increasing the exchange surface and distributing the fluids alternately into a coarse first-level checkerboard (the section of the wall is a Hilbert/Peano curve at the first iteration); subsequently, the same process is iterated twice until it obtains the finest checkerboard. The shift from one subdivision level of the Hilbert/Peano curve to the next multiplies the exchange area by 3.*

Conclusion

In this book, we have presented the formalism of boundary controlled iterated function systems (BC-IFS) as a model of geometric representation for fractal shapes.

We have introduced the concepts of this model of representation, explained their mathematical properties and given algorithms to employ them in a computer-assisted geometry design (CAGD) context. We have given numerous examples to both understand the philosophy of the model better and show its new modeling possibilities.

The strength of this model is coding the shapes using an automaton. This automaton represents the construction process of forms and directly provides an evaluation algorithm. The genericity of automatons allows for the description of complex structures. Each state of the automaton represents one type of figure and the transitions reflect how these figures are copied or included into others. The topology is then defined by the help of connections between certain selected parts. These connections take the form of equivalence relations between the finite words of the automaton. The geometry is defined using subdivision operators, namely, transformations that explain how copies of part of the form are carried out into another. The transformations then constitute the set of parameters determining the geometry associated with a topology that is determined by incidence and adjacency relations. The automaton represents the construction algorithm.

We have established a connection with NURBS and subdivision surfaces, which are the standard models for representing surfaces. Because they themselves are defined by a recursive construction process, they can be represented by a BC-IFS.

This formalism, based on automata, makes it possible to define tools for the combination of shapes, such as the tensor product of BC-IFS or the automatic generation of tree structures, whose leaves coincide with an imposed fractal surface or structure.

We have presented three examples of applications for industry: the design of lacunar structures for lightening structures and the design of rough surfaces for thermal exchange or acoustic absorption. This modeling tool is very recent. The universe of forms that it enables the generation of is huge and still unexplored. We have presented some design methodologies but many more are yet to be discovered.

Research works on this topic are still relevant, in particular, those driven by the evolution of additive manufacturing processes, always offering new possibilities. Some results could not be presented in this work, such as the control of the differential properties of fractal structures or the evaluation methods of CAD operators applied to a fractal form (Boolean operations, convex hull calculations, offset computation) or a function in general. Ongoing works focus on the use of multiscale lacunar structures for geometric and topological optimization. This problem is general and one of the current challenges of CAGD is to link the physical properties of objects to the parameters of their geometric representation.

Appendix

Data of Figures

A.1. Data of figures

A.1.1. *Example of self-similarity: Romanesco broccoli in Figure 1.10*

The seven transformations of $\mathbb{R}^3$ describing the self-similarity of Romanesco broccoli in Figure 1.10, written in Povray language:

```
#declare R1=1.;
#declare R2=-60.;

#declare s1=.5;

#declare sy=.8;
#declare sl=.8;

#declare T0= transform {
        rotate 22 * y
        scale <sl,sy,sl> // <dX, dY, dZ>
        translate <0, 0.25, 0>  // <dX, dY, dZ>
}

#declare T1= transform {
        scale s1 // <dX, dY, dZ>
        rotate R2 * z
        translate <0.5, 0, 0>  // <dX, dY, dZ>
        rotate 60 * y
}
```

```
#declare T2= transform {
        scale s1 // <dX, dY, dZ>
        rotate R2 * z
        translate <0.5, 0, 0>  // <dX, dY, dZ>
        rotate 120 * y
}
#declare T3= transform {
        scale s1 // <dX, dY, dZ>
        rotate R2 * z
        translate <0.5, 0, 0>  // <dX, dY, dZ>
        rotate 180 * y
}
#declare T4= transform {
        scale s1 // <dX, dY, dZ>
        rotate R2 * z
        translate <0.5, 0, 0>  // <dX, dY, dZ>
        rotate 240 * y


}
#declare T5= transform {
        scale s1 // <dX, dY, dZ>
        rotate R2 * z
        translate <0.5, 0, 0>  // <dX, dY, dZ>
        rotate 300 * y


}
#declare T6= transform {
        scale s1 // <dX, dY, dZ>
        rotate R2 * z
        translate <0.5, 0, 0>  // <dX, dY, dZ>
        rotate 360 * y


}
```

A.1.2. *Design example in Figure 1.48*

Left-hand side curve:

$$T_0^C = \begin{pmatrix} 1 & .5 & .2 \\ 0 & .5 & .6 \\ 0 & 0 & .2 \end{pmatrix} T_1^C = \begin{pmatrix} .2 & .5 & 0 \\ .6 & 0 & 0 \\ .2 & .5 & 1 \end{pmatrix}$$

Center curve:

$$T_0^C = \begin{pmatrix} 1 & .5 & .2 \\ 0 & 0 & .6 \\ 0 & 0.5 & .2 \end{pmatrix} \quad T_1^C = \begin{pmatrix} .2 & .5 & 0 \\ .6 & 0 & 0 \\ .2 & .5 & 1 \end{pmatrix}$$

Right-hand side curve:

$$T_0^C = \begin{pmatrix} 1 & .6 & .25 \\ 0 & .6 & .5 \\ 0 & .2 & .25 \end{pmatrix} \quad T_1^C = \begin{pmatrix} .2 & -.2 & 0 \\ .6 & .6 & 0 \\ .2 & .6 & 1 \end{pmatrix}$$

A.2. Subdivision surface in Figure 3.6

Incidence and adjacency constraints corresponding to the cellular decomposition of a quadrangular tile for a Doo-Sabin subdivision scheme.

A.2.1. *Regular case*

The constraints for the regular curve are conventional:

$$\div_0^{CR} \partial_1^{CR} \simeq \div_1^{CR} \partial_0^{CR}$$
$$\partial_0^{CR} \div_0^{SR} \simeq \div_0^{CR} \partial_0^{CR}$$
$$\partial_1^{CR} \div_0^{SR} \simeq \div_1^{CR} \partial_1^{CR}$$

The adjacency constraints on incidence operators:

$$\partial_0^R \partial_1^{CR} \simeq \partial_1^R \partial_0^{CR}$$
$$\partial_1^R \partial_1^{CR} \simeq \partial_2^R \partial_1^{CR}$$
$$\partial_2^R \partial_0^{CR} \simeq \partial_3^R \partial_1^{CR}$$
$$\partial_3^R \partial_0^{CR} \simeq \partial_0^R \partial_0^{CR}$$

Incidence constraints:

$$\div_0^R \partial_0^R \simeq \partial_0^R \div_0^{CR}$$
$$\div_1^R \partial_0^R \simeq \partial_0^R \div_1^{CR}$$
$$\div_1^R \partial_1^R \simeq \partial_1^R \div_0^{CR}$$

$$\div_2^R \partial_1^R \simeq \partial_1^R \div_1^{CR}$$
$$\div_2^R \partial_2^R \simeq \partial_2^R \div_1^{CR}$$
$$\div_3^R \partial_2^R \simeq \partial_2^R \div_0^{CR}$$
$$\div_3^R \partial_3^R \simeq \partial_3^R \div_1^{CR}$$
$$\div_0^R \partial_3^R \simeq \partial_3^R \div_0^{CR}$$

Adjacency constraints:

$$\div_0^R \partial_1^R \simeq \div_1^R \partial_3^R$$
$$\div_3^R \partial_1^R \simeq \div_2^R \partial_3^R$$
$$\div_0^R \partial_2^R \simeq \div_3^R \partial_0^R$$
$$\div_1^R \partial_2^R \simeq \div_2^R \partial_0^R$$

A.2.2. *Irregular case*

Those of the irregular curve are analogous to the regular case and inferred from the scheme of Figure 3.24:

$$\div_0^{CI} \partial_1^{CI} \simeq \div_1^{CI} \partial_0^{CR}$$
$$\partial_0^{CI} \div_0^{SI} \simeq \div_0^{CI} \partial_0^{CI}$$
$$\partial_1^{CI} \div_0^{SR} \simeq \div_1^{CI} \partial_1^{CR}$$

The adjacency constraints on the incidence operators:

$$\partial_0^I \partial_1^{CI} \simeq \partial_1^I \partial_0^{CR}$$
$$\partial_1^I \partial_1^{CR} \simeq \partial_2^I \partial_1^{CR}$$
$$\partial_2^I \partial_0^{CR} \simeq \partial_3^I \partial_1^{CI}$$
$$\partial_3^I \partial_0^{CI} \simeq \partial_0^I \partial_0^{CI}$$

Incidence constraints:

$$\div_0^I \partial_0^I \simeq \partial_0^I \div_0^{CI}$$

$$\div_1^I \partial_0^R \simeq \partial_0^I \div_1^{CI}$$

$$\div_1^I \partial_1^R \simeq \partial_1^I \div_0^{CR}$$

$$\div_2^I \partial_1^R \simeq \partial_1^I \div_1^{CR}$$

$$\div_2^I \partial_2^R \simeq \partial_2^I \div_1^{CR}$$

$$\div_3^I \partial_2^R \simeq \partial_2^I \div_0^{CR}$$

$$\div_3^I \partial_3^R \simeq \partial_3^I \div_1^{CI}$$

$$\div_0^I \partial_3^I \simeq \partial_3^I \div_0^{CI}$$

Adjacency constraints:

$$\div_0^I \partial_1^I \simeq \div_1^I \partial_3^R$$

$$\div_3^I \partial_1^R \simeq \div_2^I \partial_3^R$$

$$\div_0^I \partial_2^I \simeq \div_3^I \partial_0^R$$

$$\div_1^I \partial_2^R \simeq \div_2^I \partial_0^R$$

References

Aron, J. (2009). The Mandelbulb: First "true" 3D image of famous fractal. *New Scientist*, 204(2736), 54 [Online]. Available at: http://www.sciencedirect.com/science/article/pii/S026240790963150X.

Bandt, C. and Gummelt, P. (1997). Fractal Penrose tilings I. Construction and matching rules. *Aequationes Mathematicae*, 53(1), 295–307 [Online]. Available at: https://doi.org/10.1007/BF02215977.

Barnsley, M.F. (1986). Fractal functions and interpolation. *Constructive Approximation*, 2(1), 303–329.

Barnsley, M.F. (1988). *Fractals Everywhere*. Academic Press Professional, Inc., San Diego.

Barnsley, M.F. and Vince, A. (2013a). Developments in fractal geometry. *Bulletin of Mathematical Sciences*, 3(2), 299–348 [Online]. Available at: https://doi.org/10.1007/s13373-013-0041-3.

Barnsley, M.F. and Vince, A. (2013b). Fractal homeomorphism for bi-affine iterated function systems. *Int. J. Applied Nonlinear Science*, 1(1), 3–19.

Barnsley, M.F., Ervin, V., Hardin, D., Lancaster, J. (1986). Solution of an inverse problem for fractal and other sets. *Proceedings of the National Academy of Sciences of the United States of America*, 83, 1975–1977.

Barnsley, M.F., Hutchinson, J., Stenflo, O. (2008). V-variable fractals: Fractals with partial self similarity. *Advances in Mathematics*, 218(6), 2051–2088.

Bensoudane, H., Gentil, C., Neveu, M. (2009). Fractional half-tangent of a curve described by iterated function systems. *Journal of Applied Functional Analysis*, 4(2), 311 326.

Casteljau, P. (1985). *Mathématiques et CAO. Volume 2 : formes à pôles*. Hermes, Paris.

Chaikin, G.M. (1974). An algorithm for high-speed curve generation. *Computer Graphics and Image Processing*, 3(4), 346–349 [Online]. Available at: http://www.sciencedirect.com/science/article/pii/0146664X74900288.

Cohen, N. (1997). Fractal antenna applications in wireless telecommunications. In *Professional Program Proceedings. Electronics Industries Forum of New England*, 43–49, 6–8 May, Boston, USA.

Collet, P., Lutton, E., Raynal, F., Schoenauer, M. (2000). Polar IFS+Parisian genetic programming=Efficient IFS inverse problem solving. *Genetic Programming and Evolvable Machines*, 1(4), 339–361 [Online]. Available at: https://doi.org/10.1023/A:1010065123132.

Cox, M.G. (1972). The numerical evaluation of B-splines. *IMA Journal of Applied Mathematics*, 10(2), 134–149.

Falconer, H. (1990). *Fractal Geometry: Mathematical Foundations and Applications*, 2nd edition. Wiley, New York, USA.

Gelbrich, G. (1997). Fractal Penrose tiles ll: Tiles with fractal boundary as duals of Penrose triangles. *Aequationes Mathematicae*, 54(1), 108–116 [Online]. Available at: https://doi.org/10.1007/BF02755450.

Gouaty, G. (2009). Modélisation géométrique itérative sous contraintes. PhD thesis, École Polytechnique Fédérale de Lausanne, Lausanne, Switzerland.

Guérin, E. and Tosan, E. (2005). Fractal inverse problem: Approximation formulation and differential methods. In *Fractals in Engineering*, Lévy-Véhel, J. and Lutton, E. (eds). Springer, Heidelberg.

Hutchinson, J. (1981). Fractals and self-similarity. *Indiana Univ. Math. J.*, 30, 713–747.

Lawlor, O. (2012). GPU-accelerated rendering of unbounded nonlinear iterated function system fixed points. *ISRN Computer Graphics*, 2012.

Mauldin, R.D. and Williams, S.C. (1988). Hausdorff dimension in graph directed constructions. *Transactions of the American Mathematical Society*, 309(2), 811–829.

Mbailassem, F., Leclere, Q., Redon, E., Gourdon, E. (2018). Experimental analysis of acoustical properties of irregular cavities using laser refracto-vibrometry. *Applied Acoustics*, 130, 177–187.

Mishkinis, A., Gentil, C., Lanquetin, S., Sokolov, D. (2012). Approximate convex hull of affine iterated function system attractors. *Chaos, Solitons & Fractals*, 45, 1444–1451.

Morlet, L., Neveu, M., Lanquetin, S., Gentil, C. (2018). Barycentric combinations based subdivision shaders. *26th International Conference in Central Europe on Computer Graphics, Visualization and Computer Vision*, 49–58, 28 May–1 June, Pilsen/Prague, Czech Republic.

Morlet, L., Gentil, C., Lanquetin, S., Neveu, M., Baril, J.-L. (2019). Representation of NURBS surfaces by Controlled Iterated Functions System automata. *Computers & Graphics: X*, 2, 100006 [Online]. Available at: http://www.sciencedirect.com/science/article/pii/S2590148619300068.

Peitgen, H.-O. and Richter, P. (1986). *The Beauty of Fractals: Images of Complex Dynamical Systems*. Springer-Verlag, Heidelberg, Germany.

Pence, D. (2010). The simplicity of fractal-like flow networks for effective heat and mass transport. *Experimental Thermal and Fluid Science*, 34(4), 474–486 [Online]. Available at: http://www.sciencedirect.com/science/article/pii/S0894177709000338.

Podkorytov, S., Gentil, C., Sokolov, D., Lanquetin, S. (2012). Geometry control of the junction between two fractal curves. *Computer-Aided Design*, 42(2), 424–431.

Podkorytov, S., Gentil, C., Sokolov, D., Lanquetin, S. (2014). Joining primal/dual subdivision surfaces. In *Mathematical Methods for Curves and Surfaces*, Floater, M., Lyche, T., Mazure, M.-L., Mørken, K., Schumaker, L. (eds), Springer, Heidelberg [Online]. Available at: http://dx.doi.org/10.1007/978-3-642-54382-1_23.

Prusinkiewicz, P. and Hammel, M. (1994). *Language-restricted Iterated Function Systems, Koch Constructions, and L-systems, SIGGRAPH'94 Course Notes*. ACM Press, New York.

Prusinkiewicz, P. and Hanan, J. (1990). *Scientific Visualization and Graphics Simulation*. John Wiley & Sons, New York [Online]. Available at: http://dl.acm.org/citation.cfm?id=103356.103565.

Prusinkiewicz, P. and Lindenmayer, A. (1990). *The Algorithmic Beauty of Plants*.Springer-Verlag, Heidelberg, Germany.

Puente, C., Romeu, J., Pous, R., Garcia, X., Benitez, F. (1996). Fractal multiband antenna based on the Sierpinski gasket. *Electronics Letters*, 32(1), 1–2.

Ramshaw, L. (1989). Blossoms are polar forms. *Computer Aided Geometric Design*, 6(4), 323–358 [Online]. Available at: http://www.sciencedirect.com/science/article/pii/0167839689900320.

Requicha, A. (1996). Geometric modeling: A first course [Online]. Available at: https://www.semanticscholar.org/paper/GEOMETRIC-MODELING-%3A-A-First-Course-Requicha/191c043d2fb4786427c13ecf3fd127a83c583586.

Rian, I.M. and Sassone, M. (2014). Tree-inspired dendriforms and fractal-like branching structures in architecture: A brief historical overview. *Frontiers of Architectural Research*, 3(3), 298–323 [Online]. Available at: http://www.sciencedirect.com/science/article/pii/S2095263514000363.

Riesenfeld, R. (1975). On Chaikin's algorithm. *Computer Graphics and Image Processing*, 4(3), 304–310 [Online]. Available at: http://www.sciencedirect.com/science/article/pii/0146664X75900179.

Sapoval, B., Haeberlé, O., Russ, S. (1997). Acoustical properties of irregular and fractal cavities. *The Journal of the Acoustical Society of America*, 102(4), 2014–2019.

Sederberg, T.W., Zheng, J., Sewell, D., Sabin, M. (1998). Non-uniform recursive subdivision surfaces. In *Proceedings of the 25th Annual Conference on Computer Graphics and Interactive Techniques, ACM*, 387–394, 29–31 July, London, UK.

Sokolov, D., Gentil, C., Bensoudane, H. (2012). Differential behaviour of iteratively generated curves. In *Curves and Surfaces*, Boissonnat, J.-D., Chenin, P., Cohen, A., Gout, C., Lyche, T., Mazure, M.-L., Schumaker, L. (eds). Springer, Heidelberg [Online]. Available at: http://dx.doi.org/10.1007/978-3-642-27413-8_44.

Soo, S., Yu, K., Chiu, W. (2006). Modeling and fabrication of artistic products based on {IFS} fractal representation. *Computer-Aided Design*, 38(7), 755–769 [Online]. Available at: http://www.sciencedirect.com/science/article/pii/S0010448506000650.

Terraz, O., Guimberteau, G., Mérillou, S., Plemenos, D., Ghazanfarpour, D. (2009). 3Gmap L-systems: An application to the modelling of wood. *The Visual Computer*, 25(2), 165–180 [Online]. Available at: http://dx.doi.org/10.1007/s00371-008-0212-5.

Thollot, J. and Tosan, E. (1993). Construction of fractals using formal languages and matrices of attractors. *Proceedings of the 3rd International Conference on Computational Graphics and Visualization Techniques, Compugraphics '93*, 74–78, 5–10 December, Alvor, Portugal.

Tosan, E. (1999). Wire frame fractal topology and IFS morphisms. *4th Conference Fractals in Engineering*, 67–81, 14–16 June, Delft, The Netherlands.

Zair, C.E. and Tosan, E. (1996). Fractal modeling using free form techniques. *Computer Graphics Forum*, 15, 269–278.

Zhou, H., Sun, J., Turk, G., Rehg, J.M. (2007). Terrain synthesis from digital elevation models. *IEEE Transactions on Visualization and Computer Graphics*, 13(4), 834–848.

Index

Other titles from

in

Numerical Methods in Engineering

SIGRIST Jean-François
Numerical Simulation, An Art of Prediction 1: Theory

2017

BOROUCHAKI Houman, GEORGE Paul Louis
Meshing, Geometric Modeling and Numerical Simulation 1: Form Functions, Triangulations and Geometric Modeling
(Geometric Modeling and Applications Set – Volume 1

2016

KERN Michel
Numerical Methods for Inverse Problems

ZHANG Weihong, WAN Min
Milling Simulation: Metal Milling Mechanics, Dynamics and Clamping Principles

2015

ANDRÉ Damien, CHARLES Jean-Luc, IORDANOFF Ivan
3D Discrete Element Workbench for Highly Dynamic Thermo-mechanical Analysis
(Discrete Element Model and Simulation of Continuous Materials Behavior Set – Volume 3)

JEBAHI Mohamed, ANDRÉ Damien, TERREROS Inigo, IORDANOFF Ivan
Discrete Element Method to Model 3D Continuous Materials
(Discrete Element Model and Simulation of Continuous Materials Behavior Set – Volume 1)

JEBAHI Mohamed, DAU Frédéric, CHARLES Jean-Luc, IORDANOFF Ivan
Discrete-continuum Coupling Method to Simulate Highly Dynamic Multi-scale Problems: Simulation of Laser-induced Damage in Silica Glass
(Discrete Element Model and Simulation of Continuous Materials Behavior Set – Volume 2)

SOUZA DE CURSI Eduardo
Variational Methods for Engineers with Matlab®

2014

BECKERS Benoit, BECKERS Pierre
Reconciliation of Geometry and Perception in Radiation Physics

BERGHEAU Jean-Michel
Thermomechanical Industrial Processes: Modeling and Numerical Simulation

BONNEAU Dominique, FATU Aurelian, SOUCHET Dominique
Hydrodynamic Bearings – Volume 1
Mixed Lubrication in Hydrodynamic Bearings – Volume 2
Thermo-hydrodynamic Lubrication in Hydrodynamic Bearings – Volume 3
Internal Combustion Engine Bearings Lubrication in Hydrodynamic Bearings – Volume 4

DESCAMPS Benoît
Computational Design of Lightweight Structures: Form Finding and Optimization

2013

YASTREBOV Vladislav A.
Numerical Methods in Contact Mechanics

2012

DHATT Gouri, LEFRANÇOIS Emmanuel, TOUZOT Gilbert
Finite Element Method

SAGUET Pierre
Numerical Analysis in Electromagnetics

SAANOUNI Khemais
Damage Mechanics in Metal Forming: Advanced Modeling and Numerical Simulation

2011

CHINESTA Francisco, CESCOTTO Serge, CUETO Elias, LORONG Philippe
Natural Element Method for the Simulation of Structures and Processes

DAVIM Paulo J.
Finite Element Method in Manufacturing Processes

POMMIER Sylvie, GRAVOUIL Anthony, MOËS Nicolas, COMBESCURE Alain
Extended Finite Element Method for Crack Propagation

2010

SOUZA DE CURSI Eduardo, SAMPAIO Rubens
Modeling and Convexity

2008

BERGHEAU Jean-Michel, FORTUNIER Roland
Finite Element Simulation of Heat Transfer

EYMARD Robert
Finite Volumes for Complex Applications V: Problems and Perspectives

FREY Pascal, GEORGE Paul Louis
Mesh Generation: Application to finite elements – 2nd edition

GAY Daniel, GAMBELIN Jacques
Modeling and Dimensioning of Structures

MEUNIER Gérard
The Finite Element Method for Electromagnetic Modeling

2005

BENKHALDOUN Fayssal, OUAZAR Driss, RAGHAY Said
Finite Volumes for Complex Applications IV: Problems and Perspectives